# Untangling the Web of Hate

## Are Online "Hate Sites" Deserving of First Amendment Protection?

*Brett A. Barnett*

CAMBRIA
PRESS

Youngstown, New York

**Library of Congress Cataloging-in-Publication Data**

Barnett, Brett A.
Untangling the web of hate : are online "hate sites" deserving of First Amendment protection? / Brett A. Barnett.
p. cm.
Includes bibliographical references and index.
ISBN 978-1-934043-91-2 (alk. paper)
1. Hate speech--United States. 2. Freedom of speech—United States. 3. Internet—Moral and ethical aspects. I. Title.

KF4772.B37 2007
342.7308'53—dc22

2007035435

*To Grandma and Poppy*

# Table of Contents

# Foreword

As a communication educator and scholar, my interest in freedom of expression has been especially evident since 1981, when I began to participate in the work of the Commission on Freedom of Expression in the National Communication Association. My thinking became freshly perplexed in the latter days of the 20th century as I started to learn about public communicators' increasing use of the Internet as a means for communication. It is common knowledge that as the Internet became available to millions of people, it provided access to sending and receiving information at modest costs for many of its users. Currently, anyone who wants to be on any Internet Web site can find a spot to communicate either as an individual or as a member of an organization. Among the organizations to post their views on the Internet, of course, are those organizations that use this means of expression to espouse messages of hate. The author defines in clear and meaningful language what messages of hate are, as well as the extent to which they can be found on the Internet.

Dr. Brett Barnett has transformed his research in the areas of communication, free expression, and messages from organizations that he identifies as hate groups. He does so in a way that can engage the reader that is new to the literature on freedom of expression, and at the same time inform those who are highly familiar with the issues, cases, and complexities with which this book deals.

Few will doubt that the Internet has contributed more to the presentation and sharing of information than any other form of electronic media in our world today. As a youth growing up in Michigan during the middle part of the 20th century, the means and amounts of information available to me paled in comparison to what now lies in front of citizens in the 21st century. Moreover, the cultural norms and expectations guiding, permitting, accepting, and challenging information have dramatically changed as well. This work by Dr. Barnett takes readers into the midst of cultural changes reflected in the use of the Internet in light of current protection under the First Amendment to the U.S. Constitution. The subject, of course, is hate sites—a growing display of controversial information that, among other responses, challenges voices of moral opposition to question whether or not the continuance of such material as information should be permitted.

This book is an excellent report of an ongoing activity and controversy in America, a nation that we know guarantees freedom of expression but not without some limitations and constraints. Indeed, the U.S. Supreme Court is annually challenged with making decisions that contribute to First Amendment theory, as well as the use of information in public discourse. In the current study we find a well organized, clearly written, and carefully documented analysis of hateful expression easily located on the Internet. The author carefully constructs the background for his research and poses eight research questions to guide his effort to examine the primary interests of legal and communication scholars in their consideration of hate sites on the Internet, and whether they contain constitutionally protected content. The book is an insightful message for anyone interested in public discourse from a rhetorical, as well as a legal, point of view. The work by Dr. Barnett lays out a careful consideration of key issues grounded in the current cultural milieu. The modest amount of substantial literature that this book contributes to is strengthened by the author's work, as it partially fills an open space needing further consideration by both legal and communication scholars.

The early and well-developed chapters of this work lay out concisely the background for freedom of speech and the introduction of Internet communication in the 1990s. The questions that guided Barnett's inquiry are justifiably explained to the reader, and a pertinent literature review and explanation of both the research procedures and method of analysis are presented. Before confronting the hate sites with a keen analytical lens, the readers are taken through a carefully presented review of pertinent First Amendment cases beginning with court decisions over charges of sedition to more recent times, when the doctrine of incitement prevailed in matters where issues regarding clear and present danger were on the dockets. The "clear and present danger" doctrine is traced through key court cases leading up to what is frequently referred to as the "doctrine of incitement" that emerged from *Brandenburg v. Ohio* in 1969. By turning to key cases and decisions, this book informs us of which federal court-adopted tests have been used to make decisions about the Internet. We learn, for example, about the meaning of "true threats" in public communication as seen in the shadow of "clear and present danger" issues. Of particular importance are the arguments over the Communication Decency Act (CDA) in the Telecommunications Act of 1996. The CDA eventually was determined to be unconstitutional by the U.S. Supreme Court.

The author reminds us "by enabling virtually all persons to engage in the near-instantaneous access and dissemination of a seemingly endless array of content, the Internet has created a whole new set of challenges to free speech regulation." For the reader who is less informed, this information provides a necessary introduction to the research and findings of the author. In this book, one does not get tangled up in complexities, as the author has drawn out for us the relevant ideas, issues, and decisions that our courts have struggled with. These struggles have been made in order to maintain a fair balance of the rights and liberties in a diverse society, in accordance with legal judgments about what is allowed per the U.S. Constitution. For scholars of free expression, whether speech or print communication is a primary focus, this material is an excellent review.

While there are a substantial number of Web sites that can be considered to promote hate speech, the work done by Dr. Barnett effectively selects 10 sites representative of the more than 200 sites he visited while doing his research. In this book, he introduces us to messages from white supremacists, black separatists, and arguments of the Jewish Defense League (JDL). Religion proves to be a key theme in the rhetoric explored by the author. Evidently, the promoters of hate line themselves up with their misuse of the ideas and principles of the faithful; the Judeo-Christian faiths usually associate with the God of love. The messages from white groups that can be identified as hate speech considerably target African-Americans and Jews, and homosexuals are prime targets as well. It is argued on several of the sampled sites that blacks are inferior, as well as that Jews are taking control of the media, American businesses, and are parasitical. From the perspective of the white hate organizations, Christianity is the superior faith, while the Nation of Islam (NOI) takes an opposing view.

Symbolic expression such as music, flag burning, or images on apparel can reflect messages protected under the First Amendment. Framing his analysis using appropriate court decisions, Dr. Barnett looks at hateful rhetoric in nontextual messages on the Internet presented by the Ku Klux Klan, neo-Nazis, and the other organizations cited in his work. While Dr. Barnett clearly analyzes the nontextual rhetoric of the hate organizations, readers can easily access the sampled Web sites to confront the hate discussed by Dr. Barnett. The photographs, the computer-generated images and animations, and the sounds that are added all fit in a symbiotic way to convey hate in rhetorical forms. It is found, however, that the 10 hate groups Dr. Barnett investigated used still images as a main form of nontextual content on their sites. Within this nontextual content, the targets of hatred remain blacks, Mexicans, Jews, Muslims, and Arabs, and the distorted arguments for white Christians continue to be promoted.

As criteria for assessing the rhetoric in the hate sites, Dr. Barnett turns to four tests from constitutional cases: The *Miller* test for obscenity, the Supreme Court's "true threats" standard, the "clear and present danger" decisions, and the "bad tendency" rules. Accordingly, his findings report that the hate sites functioned with sufficient restraint to protect their expressions within the framework of our Constitution.

Are online hate sites worthy of constitutional protection under the First Amendment? Dr. Barnett's findings support the arguments that the sites and their messages, while distasteful and repugnant to most citizens, are nonetheless protected. He also challenges our thinking in relation to the future, filtering out messages, and whether or not the U.S. government may find it necessary to regulate online hate sites. For anyone interested in freedom of expression and the rhetoric found on hate sites, this book is important to read. Indeed, it provides a valuable introduction to what can be found today and what lies ahead for the messengers of hate with their words, sounds and images that continue to expand on the Internet.

John J. Makay, Ph.D.
Professor Emeritus of Communication Studies
Bowling Green State University
Ohio

# Preface

Since the research for this study was completed in 2005, the number of hate sites operated by U.S.-based hate groups has continued to escalate. At the time this study was being conducted, it was estimated that there were fewer than 500 U.S.-based hate sites (Potok, 2004). At the time this book was being written in the summer of 2007, it was estimated that the number of U.S.-based hate sites was approaching 600 (Potok, 2007). It appears that wars abroad, terrorism concerns, rising fuel prices, and immigration anxiety are helping to create an atmosphere in the United States that is increasingly uneasy, and thus particularly conducive to a proliferation of U.S.-based hate group rhetoric on the Internet.

Despite the growth in the number of U.S.-based hate sites, and the possible implications of this growth for American society and the Internet community, relatively few analyses of the content of these Web sites have been conducted. This study fills some of the void that exists regarding the analysis of the content of Web sites maintained by hate groups in the United States. Specifically, this study examined content posted on the publicly accessible areas of a sample of 10 sites maintained by organized U.S.-based hate groups.

However, unlike past studies involving analyses of hate site content, this study was not limited to white supremacist sites. The sample of hate sites examined in this study was unique in that it included sites maintained by several different kinds of hate groups: Ku Klux Klan, neo-Nazi, racist skinhead, Christian Identity, black separatist, neo-Confederate, white conservative, and pro-Jewish. The main purpose of the study was to determine whether it appeared, based on the publicly accessible content of the sampled sites, that U.S.-based hate sites are deserving of First Amendment protection.

Determining whether U.S.-based hate sites appear to be deserving of First Amendment protection is important for various reasons. One is because some groups and individuals have called for the U.S. government to ban hate Web sites, something which the governments of some less speech-protective countries already do. Although the U.S. government has not seriously considered any of these requests because of the United States' commitment to the First Amendment guarantee of free speech, the time may come when it does. Should the U.S. government ever decide to ban hate sites its decision would be made all the more easy if the Web sites did not appear to be deserving of First Amendment protection. Conversely, if hate sites did appear to be deserving of First Amendment protection, then the U.S. government would be hard-pressed to justify banning the sites.

A determination of whether Web sites operated by U.S.-based hate groups appear to be worthy of First Amendment protection is also important because neither the Supreme Court, nor any other federal courts, have made a ruling in this area. It is true that there have been federal court decisions relating to the constitutionality of hateful speech sent via e-mail and posted on antiabortion Web sites, but not Web sites operated by organized hate groups, such as the Ku Klux Klan. Yet, it is only a matter of time before a case involving a question about a hate site's constitutionality comes before a federal court. Thus, the findings of this study lend some insight into how federal courts may rule with regard to the First Amendment status of hate sites.

However, while this study's primary purpose was to determine whether it appeared that U.S.-based hate Web sites are entitled to First Amendment protection, the findings of this study make other important contributions. To begin with, this study makes an important contribution to the ongoing debate about what First Amendment tests for evaluating the constitutionality of speech are, and are not, appropriate for assessing the constitutionality of hate speech posted online. The study also revealed the various themes, targets of hatred, and nontextual rhetoric found on the sampled hate sites. Taken as whole, this information sheds light on the ideologies and underlying motivations of a variety of U.S.-based hate groups, how events and circumstances may have influenced the rhetoric these groups produce, as well as the similarities and dissimilarities that exist among the groups. Above all else, this information provides a better understanding of what type of content is posted on hate sites operated by various U.S.-based hate groups, the similarities and dissimilarities that exist in that content, as well as how that content could potentially affect persons in the online and offline communities.

Yet, even though this study is one of the most comprehensive of its kind, further analyses of the content of U.S.-based hate sites are still warranted. After all, the evidence suggests that the number of these Web sites will only continue to increase.

# Acknowledgments

I am deeply indebted to my wife, Kenna, who made it possible for me to complete this study and to get the work published. This book is the product of the sacrifices she made to help me.

I am also indebted to several of my graduate school professors for their support throughout the various phases of this project. To begin, I would like to thank Professor David Jamison, now Dean of the School of Communications and Information Systems at Robert Morris University, for the help he gave me with the legal aspects of this study while I was a master's student at the University of Akron. I would also like to thank Dr. John Makay for the considerable amount of time he devoted to helping me complete this study while I was a doctoral student at Bowling Green State University. Thanks also go to Dr. Steven Ludd, Dr. Radhika Gajjala, and Dr. Melissa Spirek for the advice and support they gave me while I was finalizing this study at Bowling Green State University.

Lastly, I would like to thank my family for making it possible for a kid from Bellville, Ohio, to go on and get a Ph.D. Without the support given to me by my family, this book would not have been possible.

# Untangling the Web of Hate

# CHAPTER 1

# INTRODUCTION

Freedom of speech is a glorious right. As Americans we routinely exercise our freedom of speech to express our views on issues ranging from the trivial to the important to the controversial. Unfortunately, however, since Americans are accustomed to being able to speak freely they often take for granted that they have the freedom of speech, forgetting that this freedom is often denied to persons in other parts of the world.

Fortunately, however, the men who drafted the Bill of Rights did not take freedom of speech for granted. After all, many of these men came to America to escape the governmental censorship that was commonplace in England at the time. The Framers of the Bill of Rights recognized that a fundamental characteristic of a free society is the ability of its citizens to exchange ideas without fear of governmental reprisal. Accordingly, the first amendment that the Framers made to the Constitution included a guarantee that Congress would "make no law...abridging the freedom of speech, or of the press."

Many scholars have observed that the Framers' First Amendment conception of free speech was based on the belief that all viewpoints should compete for acceptance in a "marketplace of ideas" free from government interference. The enduring belief in the principle of a free marketplace of ideas was demonstrated by Supreme Court Justice Oliver Wendell Holmes in his dissenting opinion in *Abrams v. United States* in 1919. In *Abrams*, Holmes wrote:

> But when men have realized that time has upset many fighting faiths, they may come to believe even more than they believe the very foundations of their own conduct that the ultimate good desired is better reached by free trade in ideas—that the best test of truth is the power of the thought to get itself accepted in the competition of the market, and that truth is the only ground upon which their wishes safely can be carried out. (*Abrams v. United States*, 1919, p. 630)

So powerful was Holmes' dissent, that some legal scholars believe, rightfully so, that it formed the foundation upon which modern First Amendment protections were built (Novick, 1989).

Seven decades after Holmes' dissent in *Abrams*, the Internet, which was commercialized in 1990, began serving as an interactive marketplace in which millions of individuals were able to communicate their views and opinions with relative ease. Today, an unprecedented number of ideas compete for acceptance on the Internet. However, the Internet is not likely the sort of free speech marketplace that Holmes or the First Amendment Framers envisioned, nor perhaps would have supported if they were alive today. While the Internet allows for heightened political and religious discussion, both at the core of the First Amendment guarantees of freedom of speech, it also permits easy access to "hate speech," content at the fringes of free speech protection.

Of course, Internet hate speech is certainly not the only type of online content to lie at the periphery of the First Amendment. However, Internet hate speech certainly does demand our attention. Hate groups such as the Ku Klux Klan are increasingly utilizing the communicative power of the Internet in an attempt to spread their message of hatred to a mass audience composed of many impressionable persons, including youth and children. Indeed, it is becoming increasingly rare for an organized hate group not to maintain a "hate site," a term commonly used to refer to an online Web site containing content that denigrates a particular class of people. Realizing this, some individuals and groups have called for the Unites States to join other countries such as France and Germany in banning hate sites. As this call has intensified, a growing number of legal scholars have explored the constitutional issues that would surround any attempt by the U.S. government to regulate Internet hate sites.

If the U.S. government ever attempts to regulate Internet hate sites, the decision to do so could not, or at least should not, occur inside a vacuum. Hate sites would need to be examined in order to determine if there was a constitutional basis for regulating their content, if not banning them altogether. This is an area in which communication scholars versed in content analysis could have conducted research that further informed the ongoing legal debate about whether hate sites should be regulated. Unlike legal scholars, however, communication scholars have not been as quick to examine the issue, let alone the content, of online hate sites. This is surprising for a variety of reasons. First, communication scholars are armed with a wealth of literature to instruct their examinations of hate site content. Moreover, as Calvert (1997) points out, hate speech is not only a "communication phenomenon" (p. 5), it also "provides an area ripe for further review by communication scholars in the empirical, cultural, and critical traditions" (p. 17). Although written in 1997, Calvert's statement rings even more true today given the ever-increasing number of hate sites on the Internet.

As a communication scholar, I recognized that there was a need to further analyze the content of Internet hate sites. Analyses of hate site content further inform the debate regarding hate Web site regulation, as well as provide both scholars and nonscholars with a better understanding of what hate sites are and what they are not. There was a particular need to further analyze the content of Web sites maintained by U.S.-based hate groups. The primary justification used for restricting freedom of expression in the United States has been the claim that certain categories of speech represent a danger to American society. If any hate sites were to pose such a significant threat to American society as to warrant their restriction, it would seem to be those maintained by U.S.-based hate groups. For one, U.S.-based hate groups have a greater likelihood of attracting American recruits than do hate groups residing in countries outside the United States. Moreover, unlike hate groups operating outside the United States, U.S.-based hate groups have the easiest physical access to American citizens, institutions, and territories should they ever feel compelled to commit any acts of violence or terrorism. While the September 11, 2001, terrorist attacks (9/11) understandably caused many in the United States to focus on the threat posed by non-Americans set on harming Americans, American society must not ignore the potential threat posed by some of its citizens within its borders.

In response to the lack of analyses of the content of U.S.-based hate sites, this study combined legal analysis with a qualitative content analysis of a sample of Web sites maintained by U.S.-based hate groups. Specifically, this study traced the historical development of the First Amendment to establish the current standards used for determining the constitutionality of both hate speech and Internet content, then examined a sample of hate Web sites to establish whether they contained constitutionally proscribable content under those standards.

# CHAPTER 2

# BACKGROUND

This chapter provides an overview of the history surrounding Internet hate sites. The chapter begins with a discussion about the evolution of U.S.-based hate sites and how legal and communication scholars have responded to the growth of these Web sites. Next, a review of the analyses of hate site content appearing in the literature of communication studies is provided. The chapter concludes with a summary of the communication literature relating to hate site content.

## THE EVOLUTION OF U.S.-BASED HATE SITES

### The 1990s: Internet Speech and Hate Speech at Issue

Once only accessible by persons working in military or research facilities, the online network that is known today as the "Internet" became available to the general public in 1990 (Lawley & Summerhill, 1993; McKay, 1996). Largely free of government regulation, the Internet allowed for robust and unfettered communication on a global scale. As improvements were made to computer technologies and the communications infrastructure, the number of persons engaging in online communication rapidly proliferated throughout the 1990s. It was not long before millions of users were online in the United States. Not surprisingly, communication scholars were soon turning their attention to Internet-related issues.

At the same time individuals were celebrating the free-expression potential of the Internet, others were voicing their concerns about the possible negative effects caused when hate speech is communicated in offline settings. Increasing debate about the wisdom of regulating hate speech would occur in the United States during the early 1990s, following an apparent rise in the amount of "hate crimes" throughout the country (Fenton, 1994; Moore, 1993). Hate crimes were so prevalent in the United States during the late 1980s that in 1990 Congress enacted the Hate Crimes Statistics Act, which required the FBI to create and maintain a system for collecting statistical data on the problem (Fenton; Jackson & Heckman, 2002). In an effort to combat the evils of bias-motivated crimes, several college campuses adopted "hate speech codes," and many cities drafted "hate crime ordinances" that included restrictions on hateful speech (Moore). Soon, legal scholars were joining the debate about hate speech, arguing both for and against its regulation. This debate would rage on throughout the 1990s after hate speech codes and hate crime ordinances started being subjected to First Amendment scrutiny by federal courts.

## Internet Speech: Communication Research and Legal Research Intersect

Given their interest in communication-related issues, communication scholars were likely the ones primarily concerned with studying the Internet as it was emerging into a mass medium during the early 1990s. However, legal scholars quickly became interested in "cyberspace" as soon as questions arose about the legality of performing transactions and posting content online. Legal scholars explored topics related to various "cybercrimes," such as electronic fraud, theft, and invasions of privacy. However, the most direct connection legal scholars made with communication scholars was through their analyses of the free speech protections that should be afforded to Internet content. As scholars in communication studies examined Internet issues and analyzed material appearing in cyberspace, legal scholars reviewed the law in the area of free expression and offered their analyses of how it applied to online content.

Legal scholars' interest in discussing online speech issues intensified after governmental concerns about the Internet's unregulated nature led Congress to enact the Telecommunications Act of 1996. The first extensive overhaul of telecommunications law since 1934, the Act included a controversial section named the Communications Decency Act (CDA) (Canis & Soriano, 1996). While the CDA addressed many worthy concerns with regard to child pornography and minors' access to obscene materials, it also had speech-restrictive provisions which many felt were both unconstitutional and ill-suited for the Internet. Immediately following the CDA's enactment, legal scholars offered their views on the Act. At the same time, a variety of groups joined the American Civil Liberties Union (ACLU) in a suit challenging the Act's constitutionality (*ACLU v. Reno*, 1996). The case of *Reno v. ACLU* (1997) would eventually reach the Supreme Court, where a majority of the justices held that the CDA's "indecent transmission" and "patently offensive display" provisions were unconstitutionally vague and placed an unacceptably heavy burden on protected speech.

However, far from ending the debate regarding the proper regulatory model for the Internet, the Court's decision left many issues of online content regulation unresolved. Since the *Reno* decision, legislators have continually failed in their efforts to draft laws in the area of online speech regulation that can withstand constitutional muster. What is more, legal scholars continue to offer a broad spectrum of opinions as to what restrictions could be placed on Internet content without violating the First Amendment.

## Internet Hate Speech: Communication Research and Legal Research Diverge

Following the *Reno* decision, Internet hate speech soon became a subject of both concern and First Amendment debate as hate groups began posting their rhetoric online. In the year preceding the passage of the Telecommunications Act of 1996 there was only one reported U.S.-based hate site, a Web site posted by former Klansman Don Black (Bostdorff, 2004; Southern Poverty Law Center, 1999). However, the year after the *Reno* decision the number of U.S.-based hate sites had escalated to 254 (Breckheimer, 2002; Southern Poverty Law Center). By the turn of the century, the number was over 400 (Southern Poverty Law Center, 2003). Some groups began calling for government officials to take steps to ban Internet hate sites. Not surprisingly, legal scholars also began participating in the hate site debate.

However, as would be expected, legal scholars examining the issue of hate sites were primarily focused on reviewing the relevant legal issues, not the actual content of these sites. Thus, the job of examining hate site content was left primarily up to communication scholars, those persons trained to analyze the content of human communications, both mediated and unmediated. Surprisingly, although communication scholars have examined the hate speech issue generally since the Internet went commercial in 1990, few articles appear in the published communication literature that deal specifically with online hate speech, and even less involve analyses of hate site content.

## LITERATURE REVIEW

Apple and Messner (2001) were among the first communication scholars to examine hate site content. The researchers analyzed the online rhetoric contained on six Web sites posted by adherents of Christian Identity, a movement that claims that whites are the superior racial group and will ultimately triumph over all other "inferior" and "evil" races in an impending race war. The researchers' rhetorical analysis demonstrated that paranoia and paradox are the essential elements of the apocalyptic rhetoric found on Christian Identity Web sites. Based on their findings, Apple and Messner concluded that "[m]ainstream America's vulnerability to hate rhetoric...may be enhanced by its dissemination via the Internet" (p. 224).

In an even more ambitious study, Bostdorff (2004) rhetorically analyzed 23 Ku Klux Klan (KKK) hate sites. She found that KKK Web sites attempt to build a community that is united by its opposition to minority groups, in particular Jews, by making appeals to white masculinity, and in some cases, segmented appeals to women, youth, and children. Bostdorff's study is one of the most in-depth and recent analyses of hate site content to appear in the communication literature. It should be noted, however, that while Bostdorff's essay was published in 2004, the hate site content she analyzed in her study was collected between November 1999 and October 2000, several months prior to 9/11.

Leets (2001), the communication scholar that has written most extensively about hate speech, also conducted an analysis of hate site content. Unlike the aforementioned content analyses, however, Leets' study was done for the express purpose of determining the constitutional protections that should be afforded to hate sites. Indeed, Leets' study stands as the first comprehensive effort to combine First Amendment analysis with an analysis of hate site content.

In her study, Leets quantitatively analyzed 266 persons' responses to hate sites. After creating a representative sample of 11 white supremacist hate sites, Leets instructed each of the volunteers in the study to select one of the sites, read it, and evaluate its content. Then, each volunteer completed a questionnaire that included a series of closed-ended questions adapted from court-adopted First Amendment formulas for determining when speech can be restricted. Results indicated that respondents felt that hate sites were low in communication value and contained indirect, insidious threats, but not advocacy for "imminent lawless action," which many believe is the current legal standard for restricting hate speech.

Based on her findings and established First Amendment guidelines, Leets drew two conclusions regarding the constitutionality of hate sites. She concluded that white supremacist sites are protected expression when viewed on the basis of short-term effects, but that the Web sites may exist only on the periphery of free speech protection if indirect or long-term effects are considered. Leets' research continues to inform the debate concerning Internet hate speech regulation and informed this study to a great degree.

## SUMMARY OF LITERATURE REVIEW

After reviewing the literature, it becomes clear that communication scholars have conducted too few analyses of hate site content, especially for the specific purpose of determining the constitutional protections that should be afforded to these controversial Web sites. Given the dangers posed by hate groups, and growing concerns about the ease with which these groups can now make their message of intolerance available to a mass Internet audience, further analysis of hate site content is warranted.

# Chapter 3

# Research Questions and Methodology

This chapter discusses the research questions that guided this study and the methodology that was used to answer those questions. The discussion relating to the research questions that guided this study has three parts. First, rationales are provided for developing the research questions. Second, the research questions are identified. Third, the relevance of the research questions to this study is discussed. The discussion relating to the methodology used in this study is divided into two parts. The first section deals with the legal analysis that was performed. The second section deals with the various procedures that were utilized during the content analysis portion of the study. The chapter concludes with a discussion about the study's overall significance.

## Research Questions

The primary purpose of this study was to answer one general question: Are online hate sites deserving of First Amendment protection?

To arrive at that answer, several specific research questions relating to First Amendment law and hate site content were developed. Answers to these questions help establish whether hate sites, in general, are entitled to First Amendment protection. Various content-related research questions were included because hate sites have rarely been content analyzed. Answers to these questions help paint a clearer picture of what hate sites are, and what they are not. In the end, eight research questions (RQ) emerged:

**RQ1**. What are the current First Amendment tests applied to hateful speech communicated in non-Internet settings?

**RQ2**. What are the current First Amendment tests applied to content posted on the Internet?

**RQ3**. To what extent are the current First Amendment tests applied to hateful speech communicated in non-Internet settings applicable to hate speech posted on the Internet?

**RQ4**. What themes appear in the rhetorical discourse found on the publicly accessible areas of Internet hate sites maintained by organized U.S.-based hate groups?

**RQ5**. What persons are the targets of hatred on the publicly accessible areas of Internet hate sites maintained by organized U.S.-based hate groups?

**RQ6**. What forms of nontextual content do organized U.S.-based hate groups post on the publicly accessible areas of their Internet hate sites?

**RQ7**. Is the content generally found on the publicly accessible areas of Internet hate sites maintained by organized U.S.-based hate groups constitutionally proscribable under the current First Amendment tests applied to hateful speech communicated in non-Internet settings?

**RQ8**. Is the content generally found on the publicly accessible areas of Internet hate sites maintained by organized U.S.-based hate groups constitutionally proscribable under the current First Amendment tests applied to content posted on the Internet?

The first three research questions represent the law review component of this study. These questions were posed in order to identify the various legal precedents relevant to the issue of online hate speech regulation. Answers to these questions established the legal foundation for assessing the constitutionality of the hate sites examined in this study.

RQ1 was posed in order to identify the various standards that the federal courts have established for determining when offline expressions of hatred for a class of people—whether spoken, put into print, or communicated symbolically—can be regulated without violating the First Amendment. These tests serve as an appropriate starting point for assessing the constitutionality of hate speech posted online. Hate speech determined to be constitutional when communicated in non-Internet settings should also be held constitutionally permissible when posted on the Internet. Moreover, at the time this study was conducted there were no federal court decisions involving mass-mediated hate speech of the sort defined in the study. Thus, the standards for non-Internet settings represented the best available legal precedents for assessing the constitutionality of hate speech posted on the sites examined in this study.

RQ2 was posed in order to reveal the various tests adopted by the federal courts for determining when textual, visual, or audio-based content posted online can be regulated without violating the First Amendment. As with the tests involved in RQ1, these standards represent the established constitutional parameters that a researcher must first consider before beginning to evaluate the constitutionality of content posted on hate sites. The constitutional protections afforded to content posted elsewhere on the Internet should apply with equal force to similar content appearing on hate sites.

Although current First Amendment tests applied to hate speech communicated in non-Internet settings offer a primary basis for assessing the constitutionality of hate speech posted online, as constitutional standards for regulating offline communications they are not necessarily applicable to Internet content. Thus, answers to RQ3 provided a means of assessing the extent to which these standards are applicable to hate speech posted on the Internet. These answers were determined in two ways. First, the commentaries of legal scholars were reviewed. Second, a review was done of the rationales for restricting speech offered by the federal courts in precedent-setting cases involving First Amendment challenges to both the regulation of non-Internet communications of hatred and Internet content in general. Determining the extent to which the rationales used to justify regulation of hateful speech in non-Internet settings are applicable to hate speech posted online provided a much more accurate basis for evaluating the constitutionality of the sampled hate sites.

RQ4, 5, and 6 represent the content analysis component of this study. These content analyses provided the evidence required to assess the constitutionality of the hate sites examined in this study.

RQ4 was posed in order to determine what themes emerged from the rhetorical discourse posted on hate sites. This data was important in assessing the constitutionality of the sampled hate Web sites specifically, and hate sites generally, since the level of protection afforded to content is largely dependent on what category of speech it is found to belong. Realizing this, many hate groups have claimed that they discuss issues of political or religious significance, both of which the courts have determined to be high-value speech deserving of the strongest First Amendment protections. An examination of the thematic content of the sampled hate sites revealed the degree of political and religious commentary found on them, and thus whether the sites should be afforded high-value speech status.

Answers to RQ5 revealed the various persons and groups of persons that U.S.-based hate organizations direct hatred against on their hate sites. Proponents of hate speech regulation have tended to argue that such expression deserves extremely limited, if any, First Amendment protection because it involves communications that are either defamatory, like notions of racial inferiority, or demeaning to "historically oppressed" groups in America. Of course, it is true that defamatory statements generally do not enjoy any First Amendment protection when they are made in reference to particular individuals or groups that are small enough that each member belonging to them could be affected as individuals (Overbeck, 2002). However, allegedly defamatory statements made about large groups of people sharing an innate human characteristic (e.g., race, sexual orientation) are entitled to significant, if not absolute, First Amendment protection (Overbeck). Thus, determining the extent to which the commentary appearing on the sampled hate Web sites involved groups, as opposed to individuals, was important in assessing the constitutionality of those sites, and hate sites in general.

This examination was also important in revealing whether the demeaning commentary posted on hate sites was confined to "historically oppressed" groups in the United States, thus bolstering the other major argument advanced by proponents of hate speech regulation. Blacks and Jews have traditionally been the primary targets of U.S.-based hate groups, which are predominantly white. However, the list of persons to which hatred is directed toward on hate sites may have expanded, and emphases changed, radically in the wake of 9/11.

Answers to RQ6 established the nontextual modes of presentation, visual or audio-based, that organized U.S.-based hate groups utilize on their Web sites. Although the First Amendment only includes explicit references to the freedoms "of *speech*" and "of the *press*," it has been interpreted as encompassing all symbolic "*expression*." Thus, while the majority of the rhetoric posted online is in the form of text (i.e., typewritten), one must consider all the forms of communication present on hate sites in assessing whether free speech protection should be afforded to them.

Research questions 7 and 8 were posed so that legal analysis could be combined with content analysis in order to determine whether the sampled hate sites were constitutionally proscribable. Taking the current First Amendment standards used to assess both hateful speech and online content revealed during the law review, and applying them to the information discovered during the evidentiary content analysis, the study culminates in legal arguments about the constitutionality of hate sites. In essence, this section of the study functions as a sort of sentencing phase in which the rule of law is applied to the evidence in order to determine the proper sentence that should be imposed.

## METHODOLOGY

The purpose of this study was to determine whether it appears that Internet hate sites, in general, are deserving of First Amendment protection. To help answer that question, a sample of Web sites maintained by U.S.-based hate groups was examined to ascertain whether they contained any constitutionally proscribable content. In order to make this determination, two analyses were performed: legal analysis and content analysis.

## LEGAL ANALYSIS

### Research Artifacts and Procedures

Two methods were used for the legal analysis components of this study: historical and categorical balancing. A historical method was used during the law review in order to establish the various legal standards under which online hate speech regulation can be viewed. This was achieved by examining the highest federal court decisions in cases involving First Amendment issues related to the regulation of expression in non-Internet settings and on the Internet. The population of applicable cases was identified by using the "Legal Research" section of the LexisNexis "Academic Search Page" to search for free-expression decisions filed since 1919, the first year in which a restriction on speech faced judicial scrutiny at the federal level. However, only those free-expression decisions that were considered most relevant to the issue of regulating hate sites were examined. For example, excluded from the analysis were any cases involving issues of media ownership or mandated technical specifications that only indirectly related to freedom of expression. This procedure still ensured that the sample of First Amendment cases reviewed in this study was sufficiently large.

Federal court decisions were chosen because they are more precedent-setting than those handed down by state courts. The decisions of the federal courts (i.e., U.S. District, U.S. Circuit, U.S. Supreme) serve as the controlling opinion in larger geographic areas. Analysis was restricted to the highest federal court decision, the most precedent-setting ruling, in a particular case. All federal courts were included in this analysis because the highest federal appeals court, the U.S. Supreme Court, issues formal signed opinions in relatively few cases per year (Overbeck, 2002), let alone cases involving the relatively new topic of Internet content regulation.

A categorical balancing method was utilized to assess the level of First Amendment protection that should be afforded to the sampled hate sites. Along with ad hoc balancing, the categorical balancing approach has been embraced by the Supreme Court (Reed, 2000). Under an ad hoc balancing approach, the Court weighs the value of free expression against other competing interests in a particular case (Reed). Reed explains that, conversely, under the categorical balancing approach:

> [T]he Court balances speech value against competing interests only in an initial case that defines a category of speech which is constitutionally unprotected under a rule asserted at the conclusion of the balancing. The rule is then applied to subsequent cases in which no ad hoc balancing is necessary because the weighing of values has occurred in the prior case establishing the category. Consequently, jurists confronting important First Amendment cases must determine whether speech falls into a previously identified, unprotected category, but need not "value" that particular speech in light of the specific facts of each case. (p. 185)

Categorical balancing was considered the most useful and appropriate method for analyzing the constitutionality of the hate sites included in this study. To begin with, the approach allowed for easier—and potentially more accurate—application of the current standards for restricting free speech revealed during the law review section of the study, which included some of the established categories of unprotected speech. Moreover, since nonjudicial categorical balancing only involves determining whether content falls into an existing class of unprotected speech, the reliance on the approach reduced the likelihood that personal value judgments about the content factored into the analysis regarding the constitutionality of the Web sites. As with jurists charged with determining whether First Amendment protection should be extended to speech, there was no need to evaluate the social worth of any specific hate site content, but only to determine the category of speech to which the content belonged.

## CONTENT ANALYSIS

After the review of the law was complete, a qualitative approach was used to examine the content found on the sampled hate sites. Analysis was restricted to hate speech posted (i.e., not sent to unwilling recipients) online, since communications of hatred sent to specific persons in a disparaged group would most likely be regarded as constitutionally proscribable forms of harassment or intimidation. What is more, unlike individually delivered messages of hatred, hate sites are readily available to a mass audience, making them the most pervasive form of hate speech on the Internet, and thus an area particularly worthy of investigation.

### Definitions

"Hate speech" is generally defined as any behavior, physical or verbal, that "arises from hostile, discriminatory, and prejudicial attitudes toward another person's [or group's] innate characteristics: sex, race, ethnicity, religion, or sexual orientation" (Epstein & Walker, 2004, p. 278). However, for the purposes of this study, a "hate site" was operationally defined simply as:

> An Internet Web site maintained by an organized hate group residing in the United States on which hatred is expressed, through any form of textual, visual, or audio-based rhetoric, for

> a person or persons, or which provides information about how individuals can support the group's ideological objectives.

By simply using the phrase "person or persons," this definition encompassed hates sites on which hatred was expressed for individuals for reasons other than their sex, race, ethnicity, religion, or sexual orientation. Given its viewpoint neutrality, the aforementioned definition encompassed not only white supremacist Web sites, but also hate sites maintained by groups whose ideologies were not white supremacist in nature. For example, the Web site maintained by the Nation of Islam, a black separatist group known for offering hate-laden commentary about Jews and whites, fell squarely under the rubric of the previously provided definition.

The use of an all-encompassing "hate site" definition was necessary to assess the complete impact that an online hate speech regulation would have on free expression on the Internet. While the online rhetoric of white supremacists may serve as the catalyst for the regulation of hate sites, any such restriction is unlikely to be limited to speech produced by these groups. History has shown that, while any free speech restriction violates the letter of the First Amendment, the courts consider subject-based regulations (e.g., no racially based hate speech) less egregious violations of the spirit of free speech than those restrictions which are based solely on a specific viewpoint (e.g., no white superiority speech). As Supreme Court Justice Stevens explained in his concurring opinion in *R.A.V. v. St. Paul* (1992), the high court has "implicitly distinguished between restrictions on expression based on *subject matter* and restrictions based on *viewpoint*, indicating that the latter are particularly pernicious [italics in original]" (p. 430). In other words, the Supreme Court has held that the First Amendment generally requires that all persons, regardless of their views, be treated equally when it comes to restricting certain categories of speech.

While on the subject of definitions, it is worth noting that, although some persons use the terms "page" and "site" interchangeably when discussing online content, in this study the term "page" is used in reference to a single Web page and "site" when referring to a collection of Web pages. Just as a home can contain several rooms, a Web site can consist of multiple Web pages. Of course, it is conceivable that some hate sites may only contain a single "hate page," a term commonly used to refer to a Web page posted on a hate Web site.

**Population of Hate Sites**

The population of hate sites was acquired from the spring 2004 edition of the *Intelligence Report* (Potok, 2004) published by the Southern Poverty Law Center (SPLC), an organization which monitors hate group activity in the United States. Each year, the SPLC identifies all of the Web sites maintained by active U.S.-based hate groups the previous year and lists them in the "Active Hate Websites" section of its *Intelligence Report.* An examination of the SPLC's "Active Hate Websites" listing for 2003 revealed that there were 497 hate sites maintained by hatemongers situated in various locations throughout the United States, including the nation's capital and all but seven states (Potok).

The SPLC was considered to be an appropriate information source for the purposes of this study for various reasons. For one, many scholars within the legal and academic communities consider the organization to be a reliable source of information about the activities of hate groups. Another reason was that other scholars doing research on hate sites have relied on the hate Web sites listing provided by the SPLC in determining what sites to content analyze. Lastly, and most importantly, the SPLC was selected because the information contained in the "Active Hate Websites" section of its annual *Intelligence Report* met virtually all the criteria contained in the definition of "hate site" used in this study. Not only was the SPLC's data restricted to U.S.-based hatemongers, its definition of what constitutes a "hate group" was not confined to white supremacist organizations. Like the definition of "hate site" used in this study, the SPLC's definition of "hate group" also encompassed black separatists and a variety of "other" organizations, such as Jewish hate groups, which were posting hateful communications online. Using the SPLC's data regarding hate groups thus enabled this study to have the necessary research focus for determining whether First Amendment protections should be extended to hate sites.

**Sampling Procedure**

Once the population of hate sites was identified, a multipart sampling procedure was applied to the large list of Web sites to create a manageable but representative sample for content analysis. The sampling process began on November 16, 2004, and was completed on November 18, 2004.

First, the Internet addresses for the sites' home pages were entered into a Web browser to determine which sites were still operational. After implementing this procedure, 191 of the sites that the SPLC identified as being active in 2003 were found to be no longer operational, further evidence of the impermanence of Web sites (Leets, 2001). Thus, the sample of hate sites was reduced to 306. All Web site addresses were double-checked to make sure that they were entered accurately before being eliminated. Interestingly, upon entering the addresses for some of the defunct Web sites, one was redirected to another hate site, usually operated by a group with a similar ideology.

After the existence of hate sites had been verified, a procedure was conducted in order to reduce analytic redundancy. Many of the hate groups identified by the SPLC appeared to simply be local chapters of a national or state hate organization. For example, while the SPLC identified various Web sites maintained by chapters of the League of the South, a "neo-Confederate" group found in several different states, the organization's national headquarters are in Killen, Alabama (Potok, 2004). If the groups are affiliated then they share a common ideology, and accordingly the content of their Web sites should be relatively similar. Therefore, in those cases where it could be determined that there were hate sites being maintained by multiple groups with the same organizational affiliation, the groups were listed as one entity and only the site operated by the head organization was included in the sample. This determination was made by examining the sites to find out whether would-be members were directed—through Internet link or by the provision of other contact information—to another chapter of the hate organization. As an example, the membership application provided by the Texas chapter for the League of the South had a Killen, Alabama address (Texas League of the South), indicating that Killen is where the organization is headquartered. If it could not be determined that a group was part of a larger hate organization, or if the Web site of its governing body could not be found, its hate site was listed individually along with those of other unaffiliated organized hate groups. In some cases, although there were several different Web site addresses listed for a group, each address connected to the same site on the Internet.

Thus, the second sampling procedure served to reduce much of the redundancy that would have been created by content analyzing a large number of Internet sites maintained by groups with a common organizational affiliation, ideology, and most likely, rhetoric of hate. This procedure further reduced the sample size by 96 to 210.

Another procedure was implemented to ensure that the sample of Internet sites to be content analyzed included only those posted on behalf of hate groups with an offline presence. Excluded from the sample were sites that the SPLC had not identified as being linked to an actual group that actively promoted hate beyond the mere posting of Internet content. Upon instituting the third sampling procedure, the overall sample size was reduced by 89 and 121 hate sites remained.

However, as SPLC's "Active Hate Websites" listing included commercial organizations, such as book publishers and music distributors, it was considered necessary to implement one final sampling procedure on the sites that still remained. Although defined as "hate groups" by the SPLC, these organizations' offline presence was almost certainly limited to the promotion and sale of hateful materials (e.g., racist music). In their study of Christian Identity Web sites, Apple and Messner (2001) excluded several sites that were listed as "hate websites" after discovering that the sites appeared to be nothing more than online clearinghouses for Christian Identity literature, operated by individuals unaffiliated with the movement. So as to avoid these same pitfalls and to exclude hate sites that may have been posted by individual hatemongers, the sample was only allowed to include Web sites that were clearly being maintained by an organized hate group. A hate group was considered "organized" if its Web site contained information regarding: (1) the group's ideology; (2) where the group was situated (i.e., city and state) in the United States; (3) how persons could contact (Internet or non-Internet) the group; (4) the group's in-person (e.g., mailings, renting billboards) activities on public properties (i.e., roads, sidewalks, courthouses) in the United States, and; (5) how persons could become a member or support the organization (e.g., donations, purchases, disseminating literature).

While any individual can set up a hate site, there were two reasons for focusing on only those posted by organized hate groups: (1) Persons searching for the type of hate speech described in this study's definition of "hate site" tend to gravitate more toward the rhetoric of organized hate groups, hatemongers that are well-known. Organized hate groups are well-known because they usually have already achieved notoriety due to their long histories or well-publicized activities, some of which are socially disruptive (or even illegal). (2) Since organized hate groups actively solicit support for their activities in the offline world (e.g., inflammatory public rallies and marches), the negative implications of their Internet communications for society are potentially higher, and thus more worthy of study.

After implementing the final sampling procedure, a total of 10 hate Web sites remained. Many of the Web sites removed during the procedure were operated by individual hatemongers or were nothing more than sites where individuals could access or purchase some form of hate speech, such as racist jokes or white supremacist music.

In some cases, Web sites operated by a head organization in a group with multiple hate sites were removed. For example, nothing was discovered on the Web site maintained by the National Alliance's headquarters to indicate that the neo-Nazi group, which had a network of hate sites at the time this sampling procedure was implemented, engaged in any in-person activities on public properties. If National Alliance was not engaging in any in-person public activities at the time it may have been largely due to the fact that the group lost nearly half of its members following the death of its founder, William Pierce, on July 23, 2002 (Potok, 2004). With such a reduction in its membership, National Alliance may no longer have had the resources to engage in much, if any, offline activity. If other hate groups also experienced significant reductions in membership, this could explain why many of the other sites in the original population of hate Web sites did not appear to be operated by an "organized hate group" as defined in this study.

**Research Artifacts**

While the multipart sampling procedure resulted in the elimination of several hundred Web sites, the remaining 10 hate sites represented a large enough sample for the purposes of this mainly exploratory study. What is more, at least one representative site remained from each of the hate Web site categories identified by the SPLC: "Ku Klux Klan," "neo-Nazi," "racist skinhead," "neo-Confederate," "Christian Identity," "black separatist," and "other."

A single Web site remained in each of the following hate Web site categories: "neo-Nazi," "racist skinhead," "neo-Confederate," and "Christian Identity." In the "neo-Nazi" category was the site maintained by the National Socialist Movement (NSM), which has its headquarters in Minneapolis, Minnesota (National Socialist Movement [NSM], n.d.). The Web site operated by the Portland-based Tualatin Valley Skins (TVS) represented the hate site in the "racist skinhead" category (Tualatin Valley Skins [TVS], n.d.). Although regarded as a "racist skinhead" group at the time of this study, TVS has since merged with the neo-Nazi NSM and is now known as "National Socialist Movement Northwest." While TVS no longer exists, the group's Web site address was still being used by National Socialist Movement Northwest at the time this book was written. In the "neo-Confederate" category was the hate site maintained by the League of the South (LOS), an organization headquartered in Killen,

Alabama (League of the South [LOS], n.d.). "Neo-Confederate" is the name that has been given to the burgeoning movement to revive the principles, including racist doctrines, of the Confederacy (Potok, 2006). Representing the "Christian Identity" category was the Web site operated by Orange County Assembly of Christ (OCAC), a Christian Identity church that was located in Brea, California at the time of this study (Orange County Assembly of Christ [OCAC], n.d.). However, it appears that OCAC is no more. At the time this book was being written the church's Web site was nonoperational and the SPLC was no longer including the organization in its listing of Christian Identity groups operating in the United States (Potok, 2007). As previously noted, Christian Identity is a fundamentally racist religion that claims that whites are superior to all other racial groups.

As a result of the multipart sampling procedure used in this study, two Web sites remained in each of the following hate Web site categories: "Ku Klux Klan," "black separatist," and "other." The hate sites in the "Ku Klux Klan" category included a Web site maintained by the American White Knights of the Ku Klux Klan (AWKKKK) and another operated by the Mississippi White Knights of the Ku Klux Klan (MWKKKK). AWKKKK is based in Cordele, Georgia (American White Knights of the Ku Klux Klan [AWKKKK], n.d.) and MWKKKK is located in Petal, Mississippi (Mississippi White Knights of the Ku Klux Klan [MWKKKK], n.d.). Web sites operated by the Nation of Islam (NOI) and the New Black Panther Party (NBPP) made up the two hate sites in the "black separatist" category. NOI is based in Chicago (Nation of Islam [NOI], n.d.), while NBPP is headquartered in Washington D.C. (New Black Panther Party [NBPP], n.d.). The two hate sites from the "other" hate group category consisted of a Web site operated by a group of white conservatives, the Council of Conservative Citizens (CCC), and another maintained by a Jewish organization, the Jewish Defense League (JDL). CCC is based out of St. Louis (Council of Conservative Citizens [CCC], n.d.), while JDL has its headquarters in Los Angeles (Jewish Defense League [JDL], n.d.).

The process of collecting content from the 10 Web sites in the final sample began on November 18, 2004, and ended on December 25, 2004. This time lag was mainly due to the large volume of content present on some sites.

Analysis was not limited to content located on the primary "home pages," but instead was conducted on all hate-group-maintained, publicly accessible (i.e., not requiring the user to enter data to gain access) content found on the Web sites included in the final sample. For example, Web pages which contained a group's hate speech were examined along with any pages containing information about how persons could assist in the furtherance of that speech (pages which were often separately linked to the home page). The subordinate pages of a Web site can contain information that is as, if not more, important to an Internet researcher than that found on the main home page. Oftentimes a home page contains nothing more than a disclaimer when controversial content such as hate speech exists on a site. Analysis was limited to content accessible by all persons since it is the public communications of hate groups which are primarily at issue in the debate over online hate speech regulation.

Given that "[t]he impermanence of Web sites is well recognized" (Leets, 2001, p. 305), computer printouts were made of each of the relevant hate pages, and additional materials (e.g., music, video) were immediately downloaded and designated by origin. All printing and downloading from a particular Web site occurred within a single 24-hour data collection period so that the materials could be content analyzed later. In those cases where content (e.g., animation) could not be printed or downloaded, it was immediately examined and the findings noted.

Information was also collected about what, if any, links to extra-organizational Web pages (i.e., Web pages maintained by other persons) existed within the sampled hate sites, although content analyses of those pages were not conducted. Of course, there was always the possibility that the links were to hate pages already included in the analysis, since hate groups sometime provide connections to Internet sites maintained by other hatemongers (Siegel, 1999). Verification that a hate page was maintained intra-organizationally (i.e., by a particular group) was accomplished by examining the Internet addresses (e.g., www.k-k-k.com) affixed to the Web pages.

Because of technological advancements in Web page design, the content of hate sites can include not only text, but also audio, motion video, still images, and a variety of computer-generated animation. Nontextual content appearing on the sampled hate sites was categorized according to the way in which it was presented.

## SIGNIFICANCE

This study is significant for a variety of reasons. To begin with, a study combining legal analysis and qualitative content analysis in order to assess the constitutionality of Internet hate sites is extremely warranted because the debate surrounding this issue will only heighten with growth in these four areas: (1) Americans' use of the Internet, (2) U.S.-based hate groups' use of the Internet, (3) diversification of racial and religious groups in the United States, and (4) the occurrence and awareness of hate crimes in the United States. Given the limited amount of research involving Internet hate speech, the findings of this study contribute significantly to the ongoing legal debate about whether hate sites should be restricted. The findings of this study also make an important contribution to the ongoing debate about what the proper scheme is for dealing with Internet hate sites, and indeed all potentially harmful online content.

Additionally, this study makes a significant contribution to the limited existing literature about the content of hate Web sites, especially those operated by U.S.-based hate groups. The comprehensive nature of the content analyses performed in the study produced a wealth of information about the various themes, targets of hatred, and nontextual rhetoric found on hate sites maintained by a variety of groups. Of particular significance is that, unlike previous analyses of hate site content (Apple & Messner, 2001; Bostdorff, 2004; Leets, 2001), this study was not limited to Web sites posted by white supremacist groups. Although she restricted her study to white supremacists, Leets (2001) conceded that "the hate movement on the Internet incorporates many groups and individuals" (p. 305). Thus, this study produces a more complete picture of what groups are actually involved in the hate movement on the Internet. What is more, this study produces a more complete picture of the similarities and dissimilarities that exist in the content of hate sites maintained by groups with both similar and dissimilar ideologies.

# Chapter 4

# Standards for Regulating Hateful Speech in Non-Internet Settings

The First Amendment was established to protect four freedoms from government censorship, two of which are the freedoms of speech and of the press. However, there have been circumstances in which exceptions to this general rule have been made. This chapter provides a historical overview of the various tests developed by the Supreme Court for determining when "hateful speech" directed toward a person or group of persons in non-Internet settings can be regulated without violating the First Amendment. "Hateful speech" is defined broadly as any sort of hostile, disapproving, or antagonistic message that persons communicate to, or about, an individual or group. Thus, "hateful speech" encompasses not only "hate speech" relating to a person's or group of persons' sex, race, ethnicity, religion, or sexual orientation, but also any hostile, disapproving, or antagonistic message arising because of an individual's or group's actions or policies. Therefore, in reviewing the standards developed by the Supreme Court for regulating "hateful speech," this chapter identifies all of the potential tests under which the constitutionality of hate sites could be assessed.

## SUPREME COURT FIRST AMENDMENT TESTS

### "Clear and Present Danger" Test

The modern history of interpretation of the right of free speech began amid the controversy surrounding American participation in World War I. In 1917, just as the United States involved itself in the hostilities in Europe, Congress enacted the Espionage Act. The 1917 Espionage Act prohibited individuals from willfully causing, or attempting to cause, dissention in the U.S. military and naval forces, or to obstruct the recruitment and enlistment of American troops, when the United States was at war. The Act's constitutionality was reviewed by the Supreme Court in the landmark free speech case of *Schenck v. United States* (1919).

*Schenck* involved Espionage Act convictions against two members of the Socialist party, Charles T. Schenck, the general secretary, and Elizabeth Baer, who was on the Executive Board. Schenck and Baer had been found guilty of violating the 1917 Act by engaging in, and conspiring to engage in, the printing and distribution of leaflets advocating resistance to the military draft. However, Schenck and Baer argued that by criminalizing their antidraft advocacy, the Congress in passing the Espionage Act had curtailed freedoms of speech and press protected by the First Amendment. The defendants appealed their convictions to the Supreme Court.

More than 125 years after the First Amendment was enacted, the Supreme Court was finally faced with the task of determining whether the Constitution demanded a literal reading of "Congress shall make no law...abridging the freedom of speech, or of the press..." The justices unanimously held that it did not. In a now memorable statement, Justice Holmes, speaking for the Court, opined:

> [T]he character of every act depends upon the circumstances in which it is done.... The most stringent protection of free speech

> would not protect a man in falsely shouting fire in a theatre and causing a panic. (*Schenck v. United States*, 1919, p. 52)

In short, the Court found that the right of free speech embodied by the First Amendment was not absolute (Fenton, 1994; Overbeck, 2002). After officially acknowledging that Congress need not honor the principle of free speech absolutism when drafting legislation, the Court considered whether the defendants' acts were nevertheless protected under the First Amendment. To decide if the activities of the *Schenck* defendants could be criminalized without violating the First Amendment, Holmes developed a test for determining when speech could be regulated. In formulating the test, Holmes held that language could be regulated if it constituted a "clear and present danger" to the government of the United States. Holmes stated:

> The question in every case is whether the words used are used in such circumstances and are of such a nature as to create a clear and present danger that they will bring about the substantive evils that Congress has a right to prevent. It is a question of proximity and degree. (*Schenck v. United States*, 1919, p. 52)

After evaluating both the law and the *Schenck* defendant's actions, the Court held that their convictions did not violate the First Amendment because their speech acts represented a "clear and present danger" within the meaning of Holmes' test.

The "clear and present danger" test is important in that it provides the first 20th century attempt to define the limits of freedom of communication (Fenton, 1994). Barron and Dienes (2004) explain that although the "clear and present danger" doctrine may appear restrictive of the government's authority to limit advocacy that it considers subversive, during the test's infancy a rigorous showing of intent was not required to satisfy the test. As the law professors write, "The requirement of intent was generally satisfied by citing the speech and allowing the jury to speculate about its possible effects" (Barron & Dienes, p. 66). This jury latitude made the test an extremely subjective one. The Court would use Holmes' "clear and present danger" test to uphold Espionage Act convictions against Jacob Frohwerk (*Frohwerk v. United States*, 1919), Eugene V. Debs (*Debs v. United States*, 1919), and Jacob Abrams and others (*Abrams v. United States*, 1919) in the same year.

However, while Holmes authored the opinions for a unified Court in *Schenck*, *Frohwerk*, and *Debs*, he would write in dissent in *Abrams*. The defendants in *Abrams* had printed and distributed circulars urging individuals not to assist the American war effort during World War I because it would ultimately be used to curtail Russians' fight for freedom. Holmes argued that these activities did not represent a "clear and present danger" within the meaning of his test. Holmes reasoned that the defendants' intent to prevent a hindrance to the revolution in Russia would not necessarily interfere with the U.S.'s war efforts in Germany, and in no way represented an attack on the American form of government.

However, not only did Holmes question whether the convictions of the *Abrams* defendants were proper under his "clear and present danger" test, he seemed to also question whether the test, in its original form, was altogether appropriate. In his dissent, Holmes modified the language he used in *Schenck*, opining, "the United States constitutionally may punish speech that produces or is intended to produce a clear and imminent danger that it will bring about forthwith certain substantive evils that the United States constitutionally may seek to prevent" (*Abrams v. United States*, 1919, p. 627). Hence, Holmes altered his "clear and present danger" test somewhat by removing the word "present" and replacing it with "imminent," a more difficult burden of proof for the government to satisfy in order to justify regulating speech. Far from an isolated occurrence, Holmes would repeatedly reiterate his newly formed opinion that imminence was a more appropriate standard for evaluating speech considered to have the potential to bring about substantive evils. Holmes similarly commented that Congress could only proscribe speech that represents a "danger of immediate evil or an intent to bring it about" (p. 628), that "imminently threaten[s] immediate interference with the lawful" (p. 630), or that "makes it immediately dangerous to leave the correction of evil counsels to time" (p. 630).

Although delivered in dissent, Holmes' opinion, which Justice Brandeis concurred with, significantly affected later judicial thinking about the "clear and present danger" test. Holmes and Brandeis would continue to revise the "clear and present danger" test in some of their subsequent dissenting opinions. In the year after the *Abrams* decision, for example, the justices would also argue that other Espionage Act convictions were unconstitutional under the "clear and present danger" test (e.g., *Pierce v. United States*, 1920; *Schaefer v. United States*, 1920).

In the 32 years following Holmes' development of the "clear and present danger" standard, as many as 37 other First Amendment cases raising the issue came before the Supreme Court (*Dennis v. United States*, 1951, pp. 591–592). In 1951, in the case of *Dennis v. United States*, the Court relied on the "clear and present danger" test to determine the constitutional validity of the Alien Registration Act of 1940, better known as the "Smith Act." The Smith Act was utilized during the communist-inspired "Red Scare" of the 1940s and 1950s. The Act provided that it was unlawful to, among other deeds, organize, or conspire to organize, any group that taught, advocated, or encouraged the overthrow of any government in the United States through force or violence.

Eugene Dennis and numerous other individuals had violated the Act between 1945 and 1948 by conspiring to organize themselves as the Communist Party of the United States of America in order to teach and advocate the overthrow of the U.S. government by force and violence. The defendants claimed that the Smith Act was unconstitutional and that its conspiracy provisions, as written and applied and construed in their case, conflicted with the First and Fifth Amendments. Nevertheless, a jury found all the *Dennis* defendants guilty, and all but one of them was fined $10,000 and sentenced to five years' imprisonment. Unsuccessful in their attempt to have the convictions overturned by a federal appeals court, the defendants appealed to the Supreme Court.

With its review limited to the contention that the Smith Act's conspiracy provisions violated the First and Fifth Amendments, the Supreme Court affirmed the *Dennis* defendants' convictions. In upholding the defendants' convictions and the Smith Act's constitutionality, the Court embraced a balancing formulation of the "clear and present danger" test developed by a lower court justice, Judge Learned Hand (Barron & Dienes, 2004; Fenton, 1994). Hand's approach required weighing a person's free speech rights against the "possible" evils that could result from his or her speech (Fenton). Justice Vinson, writing for a plurality of justices, stated:

> Overthrow of the Government by force and violence is certainly a substantial enough interest for the Government to limit speech. Indeed, this is the ultimate value of any society, for if a society cannot protect its very structure from armed internal attack, it must follow that no subordinate value can be protected. (*Dennis v. United States*, 1951, p. 509)

Hence, the Court reaffirmed that an individual's speech could be restricted if it was outweighed by a compelling governmental interest in its suppression (Fenton, 1994).

However, in adopting Hand's possible-evils standard, the Court rejected as a criterion for restricting speech the probability that a substantive evil would actually be brought about. In the Court's view, given "the inflammable nature of world conditions" (*Dennis v. United States*, 1951, p. 511) at the time, the simple fact that the defendants in *Dennis* advocated the overthrow of the government by force and violence represented a substantial enough danger to the United States as to warrant restrictions on their speech. As Vinson stated, "an attempt to overthrow the Government by force, even though doomed from the outset because of inadequate numbers or power of the revolutionists, is a sufficient evil for Congress to prevent" (*Dennis v. United States*, 1951, p. 509).

The Court's ruling in *Dennis* represented a radically speech-restrictive formulation of the "clear and present danger" test, and the decision is particularly relevant to any discussion regarding the constitutionality of Internet hate sites. The fear of communism weighed heavily in the *Dennis* Court's determination that mere advocacy of violent revolution represented a sufficient danger to justify restricting speech and assembly rights. It is thus conceivable that concerns about terrorism in the post-9/11 United States may have created an atmosphere in which similar judgments are made regarding hate speech, including that posted on the Internet. In fact, Vinson even stated that the *Dennis* Court felt that it was well within the power of government to enact a statute like the Smith Act to protect itself from change brought about by "violence, revolution and terrorism" (*Dennis v. United States*, 1951, p. 501).

Although modified and reinterpreted, the "clear and present danger" test for determining when speech can be regulated has continued to influence First Amendment law. Since developing the test, the Supreme Court has established various other exceptions to First Amendment protection.

### "Bad Tendency" Test

Six years after *Schenck*, the next important development in free speech interpretation took place in *Gitlow v. New York* (Fenton, 1994; Overbeck, 2002). While *Schenck*, *Debs*, *Frohwerk*, and *Abrams* tested a Congressional statute, many of the efforts at restricting free speech were occurring at the state level. Concerned about anarchism, bolshevism, and other threats, many states had enacted "criminal syndicalism" or "criminal anarchy" statutes. The Court in *Gitlow* addressed the constitutionality of a New York criminal anarchy statute that was nearly identical to the Smith Act. The New York statute prohibited persons from advocating, advising, or teaching the duty, propriety, or necessity of overthrowing organized government by force, violence, assassination, or any other unlawful means.

Benjamin Gitlow, a member of the Left Wing Section (LWS), a faction of the Socialist Party, had been indicted under the New York statute on two counts. Gitlow was found to have arranged for the printing, publishing, and circulation of LWS's paper, *Revolutionary Age*, and for a particularly critical article that appeared in it, "The Left Wing Manifesto." Both of these writings were found at the state level to advocate, advise, and teach "the doctrine that organized government should be overthrown by force, violence and unlawful means" (*Gitlow v. New York*, 1925, p. 655). Unlike the defendants charged under the Espionage Act, Gitlow was not advocating obstruction of any war effort. Gitlow claimed that the New York statute, as written and as applied in his case, violated the Fourteenth Amendment's due process clause. Convicted of a felony and sentenced to imprisonment, Gitlow asked the Supreme Court to review his case.

In reviewing Gitlow's case, the Supreme Court acknowledged for the first time that states, like the federal government, were also obligated to respect First Amendment freedoms of speech and of the press (Fenton, 1994; Overbeck, 2002). The Court recognized that the First Amendment's guarantees of free speech and press "are among the fundamental personal rights and 'liberties' protected by the due process clause of the Fourteenth Amendment from impairment by the States" (*Gitlow v. New York*, 1925, p. 666). However, a majority of justices held that Gitlow's speech could nonetheless be constitutionally proscribed. Justice Sanford, writing for the Court, opined:

> That a State in the exercise of its police power may punish those who abuse this freedom by utterances inimical to the public welfare, tending to corrupt public morals, incite to crime, or disturb the public peace, is not open to question. (*Gitlow v. New York*, 1925, p. 667)

In affirming Gitlow's conviction and upholding the constitutionality of New York's criminal anarchy statute, Sanford attempted to form a distinction limiting the circumstances under which the "clear and present danger" test could be invoked (Barron & Dienes, 2004).

Sanford argued that the "clear and present danger" test should only be applied in those cases involving free speech restrictions under statutes in which the legislature had not specified any language that should be proscribed. Commenting on cases involving statutes merely prohibiting specified acts thought to bring about substantive evils, without specific reference to language use, Sanford wrote:

> [T]he general statement in the *Schenck* Case...that the "question in every case is whether the words are used in such circumstances and are of such a nature as to create a clear and present danger that they will bring about the substantive evils,"...was manifestly intended, as shown by the context, to apply only in cases of this class, and has no application to those like the present, where the legislative body itself has previously determined the danger of substantive evil arising from utterances of a specified character. (*Gitlow v. New York*, 1925, p. 671)

Thus, Sanford argued that since the New York statute made certain language criminal, the constitutionality of the ordinance could not be evaluated under the "clear and present danger" test.

In upholding the validity of the New York statute as applied to Gitlow, Sanford adopted a highly speech-restrictive interpretation of some of the Court's Espionage Act decisions and formulated a standard generally called the "bad tendency" test (Barron & Dienes, 2004; Brenner, 2002; Fenton, 1994). Sanford stated that the law could be constitutionally applied to a defendant's specific utterance "if its natural tendency and probable effect was to bring about the substantive evil which the legislative body might prevent" (*Gitlow v. New York*, 1925, p. 671). Unlike the "clear and present danger" standard, the "bad tendency" test did not require that the substantive evil resulting from speech be foreseeable (Fenton). As Sanford put it:

> That utterances inciting to the overthrow of organized government by unlawful means, present a sufficient danger of substantive evil to bring their punishment within the range of legislative discretion, is clear. Such utterances, by their very nature, involve danger to the public peace and to the security of the State. They threaten breaches of the peace and ultimate revolution. And the immediate danger is none the less real and substantial, because the effect of a given utterance cannot be accurately foreseen. (*Gitlow v. New York*, 1925, p. 669)

Utilizing several combustion metaphors, Sanford argued that it might be wiser to "extinguish" a "revolutionary spark" in its incipiency than to wait until it may have begun "smoldering" and "enkindled the flame or blazed" into "a sweeping and destructive conflagration" that threatens public safety (*Gitlow v. New York*, 1925, p. 669).

Justice Holmes, joined by Justice Brandeis, dissented from the Court's ruling. Holmes held that neither the paper nor the article that Gitlow helped publish and circulate represented a "clear and present danger," the standard under which he believed the writings should have been analyzed. In his concurring opinion in *Dennis v. United States* (1951), Justice Frankfurter would explain that despite being a dissent, Holmes' opinion was "treated with the respect usually accorded to a decision" (p. 541). Holmes and Brandeis would join together in making similar arguments in support of the "clear and present danger" test in later cases.

The Supreme Court used a "bad tendency" test when it addressed the constitutionality of another state sedition act two years later in *Whitney v. California* (Brenner, 2002; Fenton, 1994). In *Whitney*, the appellant, Charlotte Anita Whitney, was found to have violated the California Criminal Syndicalism Act. The Act made it a crime to organize, assist in organizing, or to be a member of a group assembled or organized for the purposes of teaching, advocating, or aiding and abetting criminal syndicalism. "Criminal syndicalism" was defined in the Act as the commission of a crime, sabotage, or any unlawful act of force, violence, or terrorism as a way of achieving change in industrial control or ownership, or effecting political change. Whitney violated the Act through her membership and active participation in the Communist Labor Party of America, an organization that was found to teach criminal syndicalism because it called for, among other actions, the overthrow of capitalist government by the working class.

Whitney claimed that the California Act was invalid for two reasons. For one, Whitney argued that the Act was indefinite, being written in vague and uncertain language. Whitney also argued that the Act was not all-embracing, as it excluded from punishment those who advocated the use of unlawful and violent methods as a means of maintaining the status quo. Whitney further claimed, as did Gitlow, that the Act as applied in her case violated the due process clause of the Fourteenth Amendment by restraining her freedoms of speech, assembly, and association. However, like Gitlow, Whitney's defense was rejected, and she was convicted of a felony and sentenced to imprisonment, prompting her to seek relief in the Supreme Court.

The Court, however, also found Whitney's arguments unconvincing and affirmed her conviction. In dismissing Whitney's contention that the California Act operated as a denial of the rights of free speech, assembly, and association granted to her under the due process clause, the Court reiterated its reasoning in *Gitlow* (Fenton, 1994). Echoing his earlier statements in the *Gitlow* case, Justice Sanford, once again writing on behalf of the Court, stated:

> [A] State in the exercise of its police power may punish those who abuse...freedom [of speech] by utterances inimical to the public welfare, tending to incite to crime, disturb the public peace, or endanger the foundations of organized government and threaten its overthrow by unlawful means... . (*Whitney v. California*, 1927, p. 371)

Sanford's comments served as a reminder of the Court's continued reliance on the "bad tendency" test in deciding cases involving issues of free speech.

*Whitney* is significant because the Court once again relied on the "bad tendency" test to determine a statute's constitutionality. The case is of added significance because of a concurrence written by Justice Brandeis, an opinion which built on Justice Holmes' dissents and greatly influenced the subsequent life of the "clear and present danger" test (Barron & Dienes, 2004). Although Brandeis, joined by Holmes, concurred with the majority's ruling on other grounds, he expressed reservations about the wisdom of the Court's utilization of the "bad tendency" test, a highly speech-restrictive standard. After all, the *Whitney* majority's application of the "bad tendency" test made mere membership in any subversive group a punishable offense (Brenner, 2002). In developing his opinion, Brandeis sought to preserve the "clear and present danger" test by reasserting the rationale Holmes articulated in *Schenck* (Fenton, 1994). However, as Holmes did in *Abrams*, Brandeis modified the test somewhat by using the term "imminent" instead of "present" (Fenton). In his concurring opinion, Brandeis wrote:

> That the necessity which is essential to a valid restriction does not exist unless speech would produce, or is intended to produce, a clear and imminent danger of some substantive evil which the State constitutionally may seek to prevent has been settled. (*Whitney v. California*, 1927, p. 373)

Thus, Brandeis argued that government could only justifiably suppress speech when there was a reasonable basis for believing that such speech would create (or was intended to create) not just a present danger, but an impending one. Like Holmes in *Abrams*, Brandeis continually reiterated his belief that government should rely on the imminence standard in assessing whether the right of free speech, or of assembly, should be restrained. Elsewhere, Brandies opined that such restrictions were warranted only when there was reason to believe that "immediate serious violence was to be expected or was advocated" (p. 376) or when the danger or evil perceived "is so imminent that it may befall before there is opportunity for full discussion" (p. 377).

Although a concurrence, Brandeis' opinion was more precedent-setting than that written on behalf of the majority. Brandeis' argument supporting the use of a "clear and imminent danger" standard of speech regulation would resonate with later justices. The standard would be relied on by the Court in making its ruling in *Brandenburg v. Ohio* (1969), a landmark racist hate speech case decided 42 years later.

Even though the cases were decided in the 1920s, *Gitlow* and *Whitney* established what is still utilized today as the basis for analyzing speech under the "bad tendency" standard (Brenner, 2002). What is more, 21st century legal scholars still mention the "bad tendency" test in their First Amendment commentaries about hate speech.

**"Fighting Words" Test**

Two decades after rendering its opinions in *Gitlow* and *Whitney*, the Supreme Court would provide the legal world with yet another test for determining when hateful speech can be constitutionally proscribed. In 1942, legal scholars saw the establishment of a new approach to First Amendment protection and exclusion, in *Chaplinsky v. New Hampshire.* This time the Supreme Court addressed the constitutionality of a New Hampshire ordinance that proscribed utterances intended to offend, deride, or annoy another person in a public place. In particular, the ordinance stated that:

> No person shall address any offensive, derisive or annoying word to any other person who is lawfully in any street or other public place, nor call him by any offensive or derisive name, nor make any noise or exclamation in his presence and hearing with intent to deride, offend or annoy him, or to prevent him from pursuing his lawful business or occupation. (*Chaplinsky v. New Hampshire*, 1942, p. 569)

The defendant in the case, Walter Chaplinsky, was convicted of violating the ordinance by referring to a city marshal as "a God damned racketeer" and a "damned Fascist." Chaplinsky made the comments during his transport to a local police station for creating public disorder while distributing Jehovah's Witnesses literature. Chaplinsky admitted to making the comments, excepting the use of the word "God." However, he argued that the statute was unconstitutional under the Fourteenth Amendment by reason of its vagueness and indefiniteness, and because it placed an excessive restraint on the freedoms of speech, press, and worship. Chaplinsky appealed his conviction to the Supreme Court.

However, the Supreme Court unanimously rejected Chaplinsky's argument and affirmed the lower court's ruling. In sustaining Chaplinsky's conviction and the validity of the New Hampshire ordinance, the Court used a categorical approach. The justices placed insulting, or "fighting," words, along with obscenity, profanity, and libel, into a "well-defined and narrowly limited" category of speech that could be constitutionally prevented and punished. Writing for the Court, Justice Murphy defined "fighting words" as words "which by their very utterance inflict injury or tend to incite an immediate breach of the peace" (*Chaplinsky v. New Hampshire*, 1942, p. 572). Murphy then offered the Court's rationale for upholding a statute proscribing such "fighting words," writing:

> It has been well observed that such utterances are no essential part of any exposition of ideas, and are of such slight social value as a step to truth that any benefit that may be derived from them is clearly outweighed by the social interest in order and morality. (*Chaplinsky v. New Hampshire*, 1942, p. 572)

Thus, the Court reasoned that, regardless of his intent, Chaplinsky's utterances were more likely to result in a fight than to contribute to the marketplace of ideas or the quest for truth. As such, the Court ruled that Chaplinsky's speech could be constitutionally proscribed.

In an attempt to guide later efforts at determining what sorts of comments could be designated as "fighting words," the *Chaplinsky* Court developed a two-part test (Moore, 1993). As suggested by Murphy, the Court held that the two-part test for determining whether expressions qualified as "fighting words" was whether they: (1) would be perceived by people of common intelligence as having the direct tendency to immediately provoke "the average addressee to fight," and (2) were voiced during a "face-to-face" encounter (Moore, p. 1273). In subsequent decisions, the Court added four more required elements to its "fighting words" test (Moore). These elements are discussed as follows.

The 1940s had yet to come to a close when the Court added a third required element for sustaining a "fighting words" conviction (Moore, 1993). In *Terminiello v. Chicago* (1949), the first Supreme Court case involving hate speech, the Court considered a conviction under a Chicago "breach of the peace" ordinance. The defendant in the case was Arthur Terminiello, a suspended Catholic priest. While appearing at an auditorium, Terminiello had delivered an anti-Communist speech, which included disparaging comments about Russians and "Communistic" Jews. Outside of the venue, nearly 1,000 protesters grew increasingly agitated, and eventually the demonstration turned violent. Although aware of the hostile environment outside, Terminiello delivered his speech to the capacity crowd at the auditorium. Terminiello's speech further inflamed the mob of protesters. Chicago authorities charged Terminiello with disorderly conduct in violation of the Chicago breach of the peace ordinance.

An Illinois trial court instructed jurors that the Chicago ordinance included speech which "stirs the public to anger, invites dispute, brings about a condition of unrest, or creates a disturbance, or... molests...inhabitants in the enjoyment of peace and quiet by arousing alarm." (*Terminiello v. Chicago*, 1949, p. 3). With those instructions, the jury convicted and fined Terminiello. Terminiello, while never challenging the instructions, consistently maintained that the statute, as applied to his speech act, violated his constitutionally protected right of freedom of speech.

Although his conviction only amounted to a $100 fine, the Supreme Court felt that Terminiello had raised a question of considerable importance and agreed to hear his appeal. In a 5-4 decision, the Court accepted Terminiello's argument, reversed his conviction, and declared the Chicago ordinance unconstitutionally overbroad as construed by the Illinois trial court (Moore, 1993). Delivering the Court's opinion, Justice Douglas explained that it was impermissible to criminalize speech simply because it solicited public dispute, stirred individuals to anger, or created a state of unrest. Then, Douglas pointed out that since it was unclear whether Terminiello's trial verdict had rested on any of those grounds, his conviction could not stand. Douglas opined:

> [The First Amendment] may indeed best serve its high purpose when it induces a condition of unrest, creates dissatisfaction with conditions as they are, or even stirs people to anger. Speech is often provocative and challenging. It may strike at prejudices and preconceptions and have profound unsettling effects as it presses for acceptance of an idea. That is why freedom of speech, though not absolute,...is nevertheless protected against censorship or punishment, unless shown likely to produce a clear and present danger of a serious substantive

> evil that rises far above public inconvenience, annoyance, or unrest… . There is no room under our Constitution for a more restrictive view. For the alternative would lead to standardization of ideas either by legislatures, courts, or dominant political or community groups. (*Terminiello v. Chicago*, 1949, pp. 4–5)

Douglas also pointed out that since the basis for the trial verdict was not specified, it was impossible to rule on the Illinois government's argument that Terminiello's conviction rested upon his use of "fighting words," which receive no constitutional protection. Douglas indicated that the Chicago ordinance was not construed narrowly to reach only "fighting words."

Thus, the Court never considered whether Terminiello's utterances constituted "fighting words." However, in declaring the Chicago ordinance unconstitutionally overbroad as interpreted by the Illinois trial court, the *Terminiello* Court added the requirement that provocative expressions be directed against a *specific individual* to justify a "fighting words" exception from First Amendment protection (Moore, 1993).

Twenty-two years after *Terminiello*, an additional element to the "fighting words" checklist would be added during the anti-Vietnam-War protests of the early 1970s (Moore, 1993). In *Cohen v. California* (1971), Paul Robert Cohen was arrested by police for knowingly wearing a jacket bearing the phrase "Fuck the Draft" while he was in the Los Angeles County Courthouse in 1968. The Los Angeles Municipal Court found Cohen guilty of disturbing the peace in violation of a California statute prohibiting the malicious and willful disturbance of the peace of any individual or neighborhood through offensive conduct. Cohen was sentenced to 30 days incarceration.

The Court of Appeal of California affirmed Cohen's conviction, holding that his conduct constituted "fighting words" (Moore, 1993). The appellate court defined "offensive conduct" as "behavior which has a tendency to provoke *others* [italics in original] to acts of violence or to in turn disturb the peace…" (*Cohen v. California*, 1971, p. 17). California's highest court declined review. So, Cohen appealed his conviction directly to the Supreme Court. Cohen claimed that the California statute, as construed by the appellate court, infringed upon his free expression rights under the First and Fourteenth Amendments.

A divided Supreme Court reversed the lower court ruling. Justice Harlan, for the Court, provided several rationales on which the majority based its decision. For one, Harlan pointed out that Cohen's conviction could not be supported on the argument that the California statute sought to preserve appropriate decorum in the courthouse in which he was arrested since the ordinance failed to sufficiently inform the ordinary person as to what constituted inappropriate courthouse conduct. Harlan also stated that this was not an obscenity case since it could not plausibly be maintained that someone viewing Cohen's message would have been erotically stimulated. Additionally, Harlan explained that Cohen's speech could not be criminalized on the state's contention that it was thrust upon a "captive audience" inside the courthouse. As Harlan put it, individuals inside the courthouse were not a captive audience because they could have effectively avoided "further bombardment of their sensibilities simply by averting their eyes" (*Cohen v. California*, 1971, p. 21). Furthermore, Harlan indicated that the Court did not believe that Cohen's antidraft message constituted "fighting words." Harlan noted that the message was clearly not directed to any specific individual, and that there was no evidence that anyone who saw the message was provoked to committing a violent act or disturbing the peace, or that Cohen had even intended such a result.

Therefore, the Court determined that the only "conduct" which California sought to punish was Cohen's choice of words in delivering his views to the public, thus violating his rights of freedom of speech under the First and Fourteenth Amendments. Harlan opined:

> [W]hile the particular four-letter word being litigated here is perhaps more distasteful than most others of its genre, it is nevertheless often true that one man's vulgarity is another's lyric. Indeed, we think it is largely because governmental officials cannot make principled distinctions in this area that the

> Constitution leaves matters of taste and style so largely to the individual. (*Cohen v. California*, 1971, p. 25)

Thus, in overturning Cohen's conviction, the Court reaffirmed the *Terminiello* holding that expressions must be directed toward a specific person to justify the "fighting words" exception (Moore, 1993). More importantly, however, the Court also established that only extremely provocative insults were punishable as "fighting words" (Moore, 1993).

In the following year, with its ruling in *Gooding v. Wilson* (1972), the Supreme Court added yet another prerequisite to the "fighting words" test (Moore, 1993). *Gooding* involved another anti-Vietnam-War demonstration. This time, however, the defendant in the case, Johnny C. Wilson, did much more than wear a jacket in protest. In August 1966 Wilson was one of several individuals blocking an entrance to a U.S. Army headquarters in Georgia in an effort to keep out arriving inductees (Moore). When police tried to remove the protesters, Wilson became violent and verbally abusive towards two of the officers, Redding and Raborn. Wilson said to Redding, "White son of a bitch, I'll kill you," and "You son of a bitch, I'll choke you to death" (*Gooding v. Wilson*, 1972, pp. 519–520). To Raborn he remarked, "You son of a bitch, if you ever put your hands on me again, I'll cut you all to pieces" (*Gooding*, pp. 519–520). As a result, a Georgia trial court convicted Wilson on two counts of violating a Georgia statute which made it a misdemeanor for anyone to use, without provocation, opprobrious or abusive language having a tendency to create a breach of the peace.

On appeal, Wilson claimed that the statute violated his rights under the First and Fourteenth Amendments by reason of its vagueness and overbreadth. However, the Supreme Court of Georgia rejected that contention and affirmed the conviction, prompting Wilson to file for a federal habeas corpus proceeding (i.e., a proceeding to release a person from unlawful incarceration). Two federal appeals courts sided with Wilson, but Georgia appealed those decisions to the U.S. Supreme Court.

The Supreme Court affirmed the decisions of the lower federal appeals courts, holding that the Georgia breach of the peace statute under which Wilson had been convicted was unconstitutionally vague and overbroad. The author of the Court's opinion, Justice Brennan, explained that by prohibiting "opprobrious" and "abusive" language, the statute went beyond the narrow definition of "fighting words" set out in *Chaplinsky,* and as a result its application to Wilson's speech swept too broadly. He wrote, "The dictionary definitions of 'opprobrious' and 'abusive' give them greater reach than 'fighting' words... . these definitions...are not 'fighting' words as *Chaplinsky* defines them" (*Gooding v. Wilson*, 1972, p. 525).

So, while the *Gooding* Court, like the *Cohen* Court, never addressed the question of whether the defendant's utterances constituted "fighting words," it did add the requirement that only unprotected expression could be criminalized under the "fighting words" test (Moore, 1993). The Court has relied on *Gooding* to nullify other state and local ordinances designed to proscribe "fighting words" and other offensive speech (Moore).

The sixth required element of the "fighting words" test was introduced in *R.A.V. v. St. Paul* (1992), the first case involving a hate crime ordinance to come before the Supreme Court (Moore, 1993). The defendant in *R.A.V.* was Robert A. Viktora. The Court used Viktora's initials because he was a juvenile when arrested (Moore). Viktora was being prosecuted, along with several other teenagers, for violating a St. Paul, Minnesota, bias-motivated crime ordinance by burning a crudely made cross on the lawn of a black family living across the street from him. The ordinance stated:

> Whoever places on public or private property a symbol, object, appellation, characterization or graffiti, including, but not limited to, a burning cross or Nazi swastika, which one knows or has reasonable grounds to know arouses anger, alarm, or resentment in others on the basis of race, color, creed, religion or gender commits disorderly conduct and shall be guilty of a misdemeanor. (*R.A.V. v. St. Paul*, 1992, p. 380)

A Minnesota trial court granted Viktora's motion to dismiss his conviction under the ordinance on the ground that the statute violated the First Amendment through its significantly overbroad and impermissibly content-based restrictions on free speech. However, the Minnesota Supreme Court reversed that decision because it found that the St. Paul ordinance only reached conduct amounting to unprotected "fighting words" that government had a compelling interest in restricting to help promote public safety and order.

On appeal, the Supreme Court unanimously reversed the lower court's ruling, holding that the St. Paul ordinance was invalid under the First Amendment (Moore, 1993). Although the justices agreed with the Minnesota Supreme Court that the ordinance only reached expressions constituting "fighting words" within the meaning of *Chaplinsky*, they held that the statute still violated the Constitution by prohibiting speech solely based on its content. In delivering the opinion of the Court, Justice Scalia wrote:

> The First Amendment does not permit St. Paul to impose special prohibitions on those speakers who express views on disfavored subjects… . St. Paul has no such authority to license one side of a debate to fight freestyle, while requiring the other to follow Marquis of Queensberry rules… . The point of the First Amendment is that majority preferences must be expressed in some fashion other than silencing speech on the basis of its content. (*R.A.V. v. St. Paul*, 1992, pp. 391–392)

So, while the Court agreed that St. Paul's interest in protecting its citizens from bias-motivated crimes was compelling, it held that the First Amendment prohibited government from trying to serve this interest by singling out speech which invokes race, color, creed, religion, or gender for special punishment as "fighting words" (Barron & Dienes, 2004; Moore, 1993). In the Court's view, such governmental action actually went beyond content discrimination, to discrimination on the basis of viewpoint.

However, Scalia wanted to make clear that while the First Amendment generally prohibited government from utilizing content-based discriminations to force particular ideas or viewpoints out of the marketplace of ideas, it did not prohibit all forms of discrimination based on content. With respect to content-based discriminations within a proscribable category of speech, for example, Scalia explained:

> When the basis for the content discrimination consists entirely of the very reason the entire class of speech at issue is proscribable, no significant danger of idea or viewpoint discrimination exists. Such a reason, having been adjudged neutral enough to support exclusion of the entire class of speech from First Amendment protection, is also neutral enough to form the basis of distinction within the class. (*R.A.V. v. St. Paul*, 1992, p. 388)

Although it did not hold that the First Amendment prohibited all content-based speech regulations, the *R.A.V.* Court's finding still means that any speech ordinance must prohibit all "fighting words," not just a subset involving certain ideas or viewpoints disfavored by government (Moore, 1993).

*R.A.V.* is especially important in any discussion of hate site regulation because, as Barron and Dienes (2004) point out, the Court avoided making a decision as to whether any law criminalizing hate speech could pass constitutional muster. Although the Court unanimously reversed the lower court's ruling and held the ordinance unconstitutional, the justices were bitterly divided as to the logic for this result, leaving the future regulation of hate speech, including online hate speech, in question (Barron & Dienes, 2004).

Like the "clear and present danger" test, the "fighting words" test for determining when speech can be regulated continues to be acknowledged by courts asked to determine the constitutionality of hateful speech. In the recent hate speech case of *Virginia v. Black* (2003), discussed later, the Supreme Court made reference to the "fighting words" test, and all but one of the "fighting words" cases mentioned in this chapter. However, while the Court spent over half a century reformulating and rearticulating the "fighting words" test, it continued to develop new standards for evaluating whether First Amendment protection should be afforded to hateful speech.

### "Group Defamation" Test

A decade after the *Chaplinsky* Court created the "fighting words" test, a possible First Amendment exception for speech defaming a group of people would be established in the case of *Beauharnais v. Illinois* (1952). Joseph Beauharnais, president of the White Circle League of America (WCLA), had violated an Illinois "group libel" law by organizing the distribution of racist lithographs, or printed leaflets. The leaflets petitioned Chicago's mayor and city council to stop the encroachment of blacks into white populated areas. The leaflets also included a WCLA membership application and a call for one million whites living in Chicago to unite in opposition to the black residents of the area. Beauharnais had accused blacks of committing various crimes in the Chicago area.

The law at issue in *Beauharnais* provided, in part, that it was illegal in Illinois for any individual or entity to:

> [M]anufacture, sell, or offer for sale, advertise or publish, present or exhibit in any public place...any lithograph... which...portrays depravity, criminality, unchastity, or lack of virtue of a class of citizens, of any race, color, creed or religion...or...exposes the citizens of any race, color, creed or religion to contempt, derision, or obloquy or which is productive of breach of the peace or riots. (*Beauharnais v. Illinois*, 1952, p. 251)

Beauharnais contended that he could not be convicted under the Illinois group libel law because the statute violated the Fourteenth Amendment by reason of its vagueness and its restriction on the freedoms of speech and of the press. A jury rejected these contentions and Beauharnais was fined $200. Other Illinois courts sustained the conviction, prompting Beauharnais to seek relief in the Supreme Court.

The Supreme Court similarly disposed of Beauharnais' contentions and affirmed his conviction, albeit by a bare majority. Justice Frankfurter, for the Court, invoked several theories as justification for upholding the Illinois law (Barron & Dienes, 2004).

First, Frankfurter relied on a theory of libel (Barron & Dienes, 2004). Frankfurter provided a history of U.S. libel law that culminated in a reference to the *Chaplinsky* Court's unanimous holding that libelous utterances were not safeguarded by the Constitution. Then, Frankfurter explained that speech libeling a group, like that defaming an individual, was unprotected by the Constitution:

> [I]f an utterance directed at an individual may be the object of criminal sanctions, we cannot deny to a State power to punish the same utterance directed at a defined group, unless...[the punishment is]...a wilful [*sic*] and purposeless restriction unrelated to the peace and well-being of the State. (*Beauharnais v. Illinois*, 1952, p. 258)

Hence, Frankfurter theorized that Beauharnais' racist speech was, in effect, group libel which was punishable as being equivalent to a verbal attack directed at each member of the black race.

Second, Frankfurter invoked a "danger theory," claiming that hate speech directed toward a defined group could produce intergroup violence (Barron & Dienes, 2004). To illustrate his point, Frankfurter cited numerous incidents in Illinois' modern history where inflammatory speech had resulted in violence between racial and religious groups. Considering Illinois' experience with speech-induced intergroup violence, Frankfurter commented that the state's legislature had "reason in seeking ways to curb false or malicious defamation of racial and religious groups, made in public places and by means calculated to have a powerful emotional impact on those to whom it was presented" (*Beauharnais v. Illinois*, 1952, p. 261).

Lastly, Frankfurter developed a sort of "human dignity theory" (Barron & Dienes, 2004). He opined:

> It would...be arrant dogmatism, quite outside the scope of our authority in passing on the powers of a State, for us to deny that the Illinois Legislature may warrantably believe that a man's job and his educational opportunities and the dignity accorded him may depend as much on the reputation of the racial and religious group to which he willy-nilly belongs,

> as on his own merits. This being so, we are precluded from saying that speech concededly punishable when immediately directed at individuals cannot be outlawed if directed at groups with whose position and esteem in society the affiliated individual may be inextricably involved. (*Beauharnais v. Illinois*, 1952, p. 263)

The "human dignity theory" comes close to the contemporary justifications put forth by proponents of laws restricting hate speech (Barron & Dienes, 2004).

Thus, the various theories invoked by Frankfurter in *Beauharnais* help make up a sort of three-prong group libel test. In developing arguments to justify hate speech laws, some First Amendment commentators have looked to the ideas that Frankfurter employed in *Beauharnais* (Barron & Dienes, 2004). Although some question its continuing validity, not only has *Beauharnais* never been overruled, the decision is still mentioned by the Court in modern-day cases (Timofeeva, 2003). Indeed, the Court mentioned *Beauharnais* in the landmark hate speech case of *R.A.V.* in 1992. In the wake of its decision in *Beauharnais*, the Court would devise two additional tests for assessing whether hateful speech could be proscribed without violating the First Amendment.

### "Imminent Lawless Action" Test

Seventeen years after *Beauharnais*, the Supreme Court established another standard, or perhaps reformulated its "clear and present danger" test, with its decision in *Brandenburg v. Ohio* (Barron & Dienes, 2004; Brenner, 2002). In *Brandenburg*, the Holmes-Brandeis imminence standard for evaluating speech was finally adopted by a majority of justices. *Brandenburg* involved an Ohio criminal syndicalism law that was very similar to the statute at issue in *Whitney*. The Ohio law made it illegal to teach or advocate, or to assemble with persons who taught or advocated, "the duty, necessity, or propriety of crime, sabotage, violence, or unlawful methods of terrorism as a means of accomplishing industrial or political reform" (*Brandenburg v. Ohio*, 1969, pp. 444–445).

The appellant in the case, Clarence Brandenburg, a KKK leader, was convicted under the 1919 law for a speech he delivered at a Klan rally on a farm in Hamilton County, Ohio. During the speech, Brandenburg criticized the U.S. government and disparaged Jews and, in particular, blacks. Brandenburg claimed the group was considering taking revenge against the U.S. government, which he accused of suppressing the white race. The Klan leader further claimed that the group was planning to march at various locations throughout the United States. Brandenburg also made several racist comments during his speech, such as, "Personally, I believe the nigger should be returned to Africa, the Jew returned to Israel" (*Brandenburg v. Ohio*, 1969, p. 447). The only persons in attendance at the rally were the participants and a television news crew from Cincinnati that attended the meeting at Brandenburg's invitation. The news crew filmed the speech and other events that took place during the rally, which included hooded figures carrying firearms and burning a cross. Segments of the film were later aired on the Cincinnati station and on a national television network.

Brandenburg was fined $1,000 and sentenced to a jail term of 1 to 10 years. Brandenburg challenged the constitutionality of the Ohio criminal syndicalism law under the First and Fourteenth Amendments and appealed his conviction all the way to the Supreme Court.

The Supreme Court unanimously sided with Brandenburg and overturned the Ohio law. The Court held in a *per curiam* (i.e., unsigned) opinion that the law, as written and as applied, violated the First and Fourteenth Amendments by punishing individuals who merely advocated, or assembled with persons who merely advocated, lawless action. In finding the Ohio criminal syndicalism law unconstitutional, the Court, citing *Dennis*, declared:

> *Whitney* has been thoroughly discredited by later decisions … . These later decisions have fashioned the principle that the constitutional guarantees of free speech and free press do not permit a State to forbid or proscribe advocacy of the use of force or of law violation except where such advocacy is directed to inciting or producing imminent lawless action

> and is likely to incite or produce such action. (*Brandenburg v. Ohio*, 1969, p. 447)

In overturning the Ohio law, and in reversing *Whitney*, the *Brandenburg* Court went beyond *Dennis* and created a new standard for assessing the constitutionality of speech now known as the "imminent lawless action" test (Brenner, 2002; Fenton, 1994). Breaking down the Court's language, one finds that the "imminent lawless action" test has three interrelated parts: (1) there must be some "advocacy of the use of force or of law violation," (2) that advocacy must have been "directed to inciting or producing imminent lawless action," and (3) that advocacy must be "likely to incite or produce such action."

Thus, the "imminent lawless action" test differs from the "bad tendency" test in that it only allows for the suppression of speech that is intended to, and likely to, create impending unlawful conduct, not the mere advocacy of such action (Fenton, 1994). Barron and Dienes (2004) also point out that the Court in *Brandenburg* demanded a greater showing "of imminent danger from incitement than Frankfurter required in *Beauharnais*" (p. 205). As previously noted, the only people attending the rally at issue in *Brandenburg* were a news crew, invited by Brandenburg, and the Klansmen themselves.

So, by demanding a greater showing of the likelihood of danger resulting from speech, the *Brandenburg* Court limited the circumstances under which an individual could be punished for his or her expressive activities (Brenner, 2002), including hate speech. In the wake of *Brandenburg*, federal courts have applied the "imminent lawless action" test in a variety of cases involving hostile statements. What is more, the "imminent lawless action" test remains the modern standard for determining when hate speech can be constitutionally regulated (Brenner, 2002; Komasara, 2002). As Salimipour (2001–2002) notes, following the *Brandenburg* decision "U.S. Courts have continually stressed the need for imminent lawless action" (p. 401).

**"True Threats" Test**

*Brandenburg* was not the only important ruling in the area of hateful speech regulation handed down by the Court in 1969. Two months before *Brandenburg*, the Supreme Court paved the way for the development of an additional standard for evaluating the constitutionality of hateful speech in *Watts v. United States* (1969), yet another landmark free speech decision involving antiwar rhetoric. At issue in the case were comments made in 1966 by the *Watts* defendant, an 18-year-old man, regarding then-President Lyndon B. Johnson. While at a politically oriented public gathering on the grounds of the Washington Monument attended by several people, including an Army Counter Intelligence Corps investigator, Watts commented:

> They always holler at us to get an education. And now I have already received my draft classification as 1-A and I have got to report for my physical this Monday coming. I am not going. If they ever make me carry a rifle the first man I want to get in my sights is L.B.J.... . They are not going to make me kill my black brothers. (*Watts*, p. 706)

After making the statement, Watts and the crowd laughed. Nevertheless, Watts was arrested for making the comments and charged with violating a 1917 statute that made it a felony to knowingly and willfully threaten to kill, or to inflict physical harm upon, the president.

Watts' defense lawyer contended that there was no basis for finding that the defendant's comments constituted a threat to the president. The lawyer pointed out that Watts' comments were both conditional in nature and made during a political debate. However, a jury in the U.S. District Court for the District of Columbia found Watts guilty of violating the 1917 statute. The U.S. Court of Appeals for the D.C. Circuit affirmed the felony conviction.

On appeal, the Supreme Court reversed Watts' conviction. The Court found that the 1917 statute was constitutional given the nation's overwhelming interest in insuring both the president's safety and his ability to perform executive duties without distraction. However, the Court held that the statute's application to Watts' utterances was invalid under the First Amendment because the government had failed to prove that his statements constituted a "true threat" within the meaning of the ordinance. The Court stated, in a *per curiam* opinion,

"What is a threat must be distinguished from what is constitutionally protected speech" (*Watts v. United States*, 1969, p. 707). Although it considered Watts' statement to be a very crude and offensive way of voicing one's political opposition to the president, the Court wrote, "The language of the political arena, like the language used in labor disputes,...is often vituperative, abusive, and inexact" (*Watts v. United States*, 1969, p. 708). In the Court's view, Watts' comments, taken in context, constituted a kind of political hyperbole rather than a "true threat" unprotected by the First Amendment.

With its decision in *Watts*, the Supreme Court laid the foundation on which modern First Amendment "true threats" analysis has been built (Brenner, 2002). However, although it stated that "true threats" are unprotected by the First Amendment, the *Watts* Court failed to provide a specific test for determining when a threat was "true" (Brenner). In the absence of any Supreme Court standard, various courts of appeal were required to draft their own "true threats" tests (Brenner). However, in *Virginia v. Black* (2003), the most recent case involving hate speech to come before the Supreme Court, the high court picked up where it left off in *Watts* and formulated a "true threats" test for assessing whether speech can be regulated consistent with the First Amendment.

Like *R.A.V.*, *Virginia v. Black* involved convictions under a statute that prohibited cross burning. Unlike the content-based ordinance struck down in *R.A.V.*, however, the Virginia statute did not impose special prohibitions on cross burners expressing views on specified disfavored subjects, such as racial or religious hatred. The Virginia statute made it a felony for anyone, regardless of his or her motivation, to burn a cross on private property or public land if done with "an intent to intimidate a person or group of persons" (*Virginia v. Black*, 2003, p. 348). However, the statute provided that cross burning, by itself, would be considered *prima facie* evidence (i.e., proof on its face) of intent to intimidate.

Three individuals, Barry Black, Richard Elliot, and Jonathan O'Mara, were convicted separately of violating the Virginia statute for cross-burning activities they had engaged in during two different incidents. Black was charged with violating the statute for his role in leading a KKK rally on private property, with the owner's permission, during which a 25- to 30-foot cross was burned within view of several private residences and a public highway. Over his objection that the Virginia statute's *prima facie* evidence provision violated the First Amendment, a jury found Black guilty and fined him $2,500. Elliot and O'Mara, who were not Klansmen, had violated the Virginia statute by partially burning a cross on the yard of Elliot's black neighbor. During separate proceedings, both men were fined $2,500 and sentenced to 90 days in jail.

All three defendants appealed to Virginia's Supreme Court, arguing that the statute under which they were convicted was unconstitutional. The Supreme Court of Virginia consolidated the cases, and, relying upon *R.A.V.* as precedent, ruled in the defendants' favor, holding that the statute was unconstitutional. This ruling prompted Virginia to appeal to the U.S. Supreme Court.

On appeal, a divided Supreme Court upheld the lower court's judgment with respect to Black, but vacated the judgment with respect to O'Mara and Elliot. While it held that the First Amendment permits content-neutral ordinances prohibiting cross burning done with the intent to intimidate, the Court found that there were problems with how the Virginia statute's *prima facie* evidence provision was interpreted in all three cases.

With regard to Black, the Court held that his conviction could not stand because the provision, as interpreted in instructions to the jury and as applied in his case, was unconstitutional on its face. The majority found that the provision violated the First Amendment by failing to distinguish between cross burning done with intent to intimidate and cross burning done for the purpose of engaging in constitutionally protected symbolic expression. Justice O'Connor, for the Court, opined, "The prima facie evidence provision in this case ignores all of the contextual factors that are necessary to decide whether a particular cross burning is intended to intimidate. The First Amendment does not permit such a shortcut" (*Virginia v. Black*, 2003, p. 367). As O'Connor explained, while burning a cross may be intended to intimidate, the same act is also sometimes performed to symbolize group solidarity, to make an ideological or political statement, or as part of film and stage productions. In short, O'Connor pointed out that "the burning of a cross is symbolic expression" (*Virginia*, p. 360) which may or may not involve constitutionally proscribable intimidation.

As for O'Mara and Elliot, the Court vacated the judgments against them because no instructions on the *prima facie* evidence provision were provided during the earlier proceedings. However, the Court left open the possibility that the lower court could constitutionally interpret, or sever, the provision. Therefore, the Court remanded the defendants' case for retrial.

Hence, the *Black* Court's ruling hinged on the Virginia statute's *prima facie* evidence provision. However, O'Connor took the opportunity to explain why a majority of the justices felt the ordinance did not violate the First Amendment insofar as it outlawed, in content-neutral fashion, cross burning intended to intimidate. Citing *Watts* and *R.A.V.*, she stated:

> [T]he First Amendment...permits a State to ban a "true threat."..."True threats" encompass those statements where the speaker means to communicate a serious expression of an intent to commit an act of unlawful violence to a particular individual or group of individuals... . The speaker need not actually intend to carry out the threat. Rather, a prohibition on true threats "protects individuals from the fear of violence" and "from the disruption that fear engenders," in addition to protecting people "from the possibility that the threatened violence will occur."... Intimidation in the constitutionally proscribable sense of the word is a type of true threat, where a speaker directs a threat to a person or group of persons with the intent of placing the victim in fear of bodily harm or death... . [S]ome cross burnings fit within this meaning of intimidating speech... . [T]he history of cross burning in this country shows that cross burning is often intimidating, intended to create a pervasive fear in victims that they are a target of violence. (*Virginia v. Black*, 2003, pp. 359–360)

Thus, whereas *Watts* established that society's overwhelming interest in safeguarding the president's ability to effectively serve permits banning only "true threats" of violence against the president, *Black* established that, because cross burning is so intimidating, states may constitutionally proscribe it when intimidation is the burner's intention, without prohibiting other intimidating messages.

O'Connor's commentary served two purposes. First, she was able to clarify that the *R.A.V.* Court held that the First Amendment prohibited some, but not all, content-based discriminations within a proscribable class of speech, an issue which some legal scholars questioned. Second, O'Connor formulated the Court's "true threats" test for evaluating hateful speech. Under the guidelines set forth by O'Connor, expressive activity may be considered a "true threat," unprotected by the First Amendment, if it is: (1) done for the purposes of communicating "a serious expression of an intent to commit an act of unlawful violence," *or* "placing [persons] in fear of" being victimized by "unlawful violence;" and (2) directed "to a *particular* individual or group of individuals." By placing speech intended to intimidate into the same category as serious expressions of intent to commit unlawful acts of violence, O'Connor expanded the field of expressive activities that may be proscribed under a "true threats" analysis.

The "true threats" test formulated by O'Connor is but the latest in the string of standards established by the Supreme Court for assessing whether constitutional protections should be extended to hateful speech. Given that legislatures continue to rely on laws prohibiting hateful speech in their efforts to protect citizens from the threats posed by bias-motivated crime and other possible evils, more standards are likely to be formulated by the Court in the future.

## Summary

The Supreme Court has developed six major First Amendment tests for determining when government may abridge individuals' freedom to deliver communications of hatred for a class of people through traditional, non-Internet modes of communication: "clear and present danger," "bad tendency," "fighting words," "group defamation," "imminent lawless action," and "true threats." Several patterns emerge with regard to these First Amendment tests.

First, all of the tests were devised during periods of extreme turbulence in America. Indeed, the oldest of these First Amendment standards, the "clear and present danger" test, was formulated in March 1919 as World War I was drawing to a close. Six years after devising the "clear and present danger" standard, the Court developed its "bad tendency" test at a time when political instability in other countries had caused heightened concerns about the overthrow of government in the United States. However, since its formulation of the "bad tendency" test, all the rest of the Court's First Amendment standards for regulating non-Internet communications of hatred have also been developed during times of war. The Court devised its test for "fighting words" during World War II, for "group defamation" while the Korean War was taking place, for "imminent lawless action" at the same time as the Vietnam War, and for "true threats" in the midst of the current "war on terrorism." If history is any indicator, it seems that the less stable the conditions are in American society, the more likely it is that the Supreme Court will develop a First Amendment test for regulating hateful speech.

A second pattern that emerges with regard to the First Amendment tests identified in this chapter pertains to the scope of their application. All of the tests provide guidelines for determining whether particular non-Internet communications of hatred are likely to produce harm sufficient to render them constitutionally proscribable. However, the content to which the standards were intended to apply has generally shifted from speech involving government to speech directed at individuals. The earliest two First Amendment standards for regulating hateful speech in non-Internet settings, the "clear and present danger" and "bad tendency" tests, were formulated to provide constitutional guidelines for proscribing speech thought likely to result in actions that would threaten government, and by extension society at large. However, excepting the "imminent lawless action test," the Court's later First Amendment tests have been more limited in their scope, applying only to speech believed liable to produce a threat to the well-being of individuals or defined groups.

Finally, one notices that there has been an increase in the degree of specificity provided by the Court's First Amendment tests for regulating communications of hatred in non-Internet settings. The "clear and present danger" and "bad tendency" standards provided a single, general guideline for evaluating the constitutionality of speech. However, the Court's later First Amendment tests required that several—often very specific—criteria be met in order to proscribe a particular communication. Of the last four multipart tests that the Court has formulated, its "true threats" standard is two-prong, its "group defamation" and "imminent lawless action" standards are three-prong, and its "fighting words" standard, while initially two-prong, has evolved into a six-part test.

## Chapter 5

# Standards for Regulating Content Posted on the Internet

In 1919, when the Supreme Court was first called upon to determine how a law abridging expression comported with the First Amendment, cases such as *Schenck* and *Debs* involved the dissemination of information merely by way of the printed and spoken word. As electronic media began to develop later in the 20th century, enabling messages to be propagated through wires, cables, and the air, it further complicated the question of how the First Amendment should be applied to communications. This is certainly true of the Internet, an interactive medium enabling all forms of human communication. By enabling virtually all persons to engage in the near-instantaneous access and dissemination of a seemingly endless array of content, the Internet has created a whole new set of challenges to free speech regulation. And, as is usually the case when new media are involved, courts have struggled with how to adjudicate Internet speech cases in accordance with established First Amendment jurisprudence. Faced with the challenge of how to properly decide cases involving the regulation of content on the Internet, courts have often looked to the First Amendment tests developed in pre-Internet (i.e., 1990) cases.

This chapter examines the various First Amendment tests adopted by the Supreme Court and other federal courts for evaluating Internet communications. Although federal court rulings involving questions about the constitutionality of materials disseminated online are few in number relative to the total number of free speech decisions, the standards adopted in the cases serve as Internet-specific tests for assessing the constitutionality of hate site content.

## FEDERAL COURT-ADOPTED TESTS FOR INTERNET CONTENT

### "True Threats" Tests Prior to *Virginia v. Black*

Almost immediately after the commercialization of the Internet, cases involving issues of online free speech started coming before federal courts (e.g., *Cubby, Inc. v. CompuServe, Inc.*, 1991). However, the first federally adjudicated Internet speech case of relevance to this study did not take place until the mid-1990s when a federal district court in Michigan heard *United States v. Baker* (1995).

*Baker* involved hateful speech about females. The defendant in the case, Abraham Jacob Alkhabaz, a University of Michigan student who went by the name "Jake Baker," was being prosecuted under a federal statute that states in part:

> Whoever transmits in interstate or foreign commerce any communication containing any threat to kidnap any person or any threat to injure the person of another, shall be fined...or imprisoned not more than five years, or both. (*United States v. Baker*, 1995, p. 1380)

Government officials claimed that Baker had violated the statute through a series of e-mails he had exchanged with "Arthur Gonda," a person believed to be living in Ontario, Canada, but whose true identity and whereabouts could not be established at trial. For approximately two months, Baker and Gonda had exchanged private e-mails in which they expressed a sexual interest in committing acts of violence against girls and women. One of the allegedly threatening communications for which Baker was charged under the statute went in part:

> One of the things I've started doing is going back and re-reading earlier messages of yours. Each time I do[ ] they turn me on more and more. I can't wait to see you in person. I've been trying to think of secluded spots[,] but my knowledge of Ann Arbor is mostly limited to the campus. I don't want any blood in my room, though I have come upon an excellent method to abduct a bitch[.]
>
> As I said before, my room is right across from the girl's bathroom. Wiat [*sic*] until late at night[,] grab her when she goes to unlock the dorr [*sic*]. Knock her unconscious[ ] and put her into one of those portable lockers (forget the word for it)[ ] or even a duffle bag. Then hurry her out to the car and take her away...What do you think? (*United States v. Baker*, 1995, p. 1388)

This statement was representative of the larger e-mail exchange between Baker and Gonda. Although no individuals were named in the Baker-Gonda communications, the government claimed that the targets of the e-mailed statements included a woman known to Baker and females residing in his dormitory, university neighborhood, and hometown in Ohio. Baker claimed, however, that the content of the e-mails he exchanged with Gonda was protected by the First Amendment.

The U.S. District Court for the Eastern District of Michigan ruled in Baker's favor, holding that the e-mails exchanged between him and Gonda constituted protected speech rather than "true threats" within the meaning of the statute. In making this determination, the District Court adopted a "true threats" test established by the Second Circuit Court of Appeals in *United States v. Kelner* (1976).

As mentioned earlier, after the Supreme Court failed to provide a "true threats" test in *Watts*, courts of appeal were required to draft their own standards for determining when communications constituted actual threats. Utilizing its "true threats" test, the *Kelner* Court upheld a defendant's conviction for making a threat during an interview that was filmed and later broadcast on television.

After applying the *Kelner* Court's "true threats" test, the District Court held that it would be unconstitutional to criminalize the Baker-Gonda e-mails as they contained no specific expressions of intent to immediately commit acts of violence. The District Court suggested that in the absence of such a threat the Baker-Gonda communications amounted to nothing more than a "rather savage and tasteless piece of fiction" (*United States v. Baker*, 1995, p. 1390).

In utilizing the *Kelner* Court's standard to overturn the indictment against Baker, the federal District Court established that a "true threats" test is applicable to hateful messages transmitted over the Internet. The *Baker* Court's conclusion has been repeatedly reinforced, starting with the Sixth Circuit Court of Appeals' review of its decision in *United States v. Alkhabaz* (1997), a case in which "Jake Baker" went by his given name.

In *Alkhabaz*, the Sixth Circuit Court held that the District Court properly concluded that the Baker-Gonda e-mails did not constitute "true threats." Since *Alkhabaz*, other federal courts have further reinforced the *Baker* Court's conclusion by also utilizing "true threats" tests to assess the constitutionality of allegedly threatening statements delivered person-to-person via the Internet. These cases fully established that a "true threats" test is applicable to private (i.e., person-to-person) online communications that are hateful in nature. However, these cases did not establish that such a standard could be applied to hateful content posted on publicly accessible areas of the Internet, such as hate sites. This latter point would be established in *Planned Parenthood of the Columbia/Williamette v. American Coalition of Life Activists* (2002), a high-profile and highly complicated case involving hateful antiabortion rhetoric.

In *Planned Parenthood of the Columbia/Williamette v. American Coalition of Life Activists* (hereinafter *PPCW*), the Ninth Circuit Court of Appeals was asked to determine whether certain antiabortion materials were protected by the First Amendment. At issue in the case were an antiabortion Web site and two "wanted"-type antiabortion posters that were circulated at events and appeared in literature published by pro-life advocates. Discussion will also be afforded to these two posters since they factored into the Circuit Court's analysis regarding the constitutionality of the antiabortion Web site at issue in *PPCW.*

Posted in January 1997, the antiabortion Web site was maintained by the American Coalition of Life Activists (ACLA), a Portland-based group. The site was a compilation about those individuals whom the ACLA anticipated might someday be tried for crimes against humanity. Accordingly, ACLA titled the Web site "Nuremberg Files," after the "Nuremberg Trials" in which Nazis were tried for crimes against humanity. The site included an "ABORTIONISTS" section, which listed approximately 200 doctors who performed abortions, as well as a file listing over 200 other individuals such as judges, politicians, and abortion rights supporters.

However, ACLA went beyond merely posting names on the "Nuremberg Files" Web site. By modifying the appearance of the doctors' names listed in the site's "ABORTIONISTS" section, ACLA also provided the physical statuses of those individuals. While a name would be entered in black font, indicating that the doctor was performing abortions, it would be greyed-out if that individual had been wounded, and struck through if he or she had been killed. ACLA's conduct in listing names on its Web site would be a central issue when the group was later prosecuted.

ACLA was also responsible for creating and circulating the two "wanted"-type antiabortion posters at issue in *PPCW.* One of the posters was unveiled at a pro-life event in Washington, D.C. in January 1995. That poster, which bore the heading "THE DEADLY DOZEN," identified 13 doctors who performed abortions, or, as ACLA referred to them, "abortionists." The poster, which was made to look like the criminal "wanted" posters used by authorities, also provided these doctors' home addresses and offered a "5,000 REWARD" "for information leading to arrest, conviction and revocation of license to practice medicine" (*PPCW*, 2002, p. 1065). In addition, "The Deadly Dozen" poster included the following antiabortion rhetoric: "Abortion was provided as a choice for East European and Jewish women by the (Nazi) National Socialist Regime, and was prosecuted during the Nuremberg Trials (1945–46) under Allied Control Order No. 10 as a 'war crime' " (*PPCW*, p. 1064). The term "ABORTIONIST," written in large, boldface type, and the word "GUILTY," followed by the phrase "OF CRIMES AGAINST HUMANITY," also appeared on the poster (*PPCW*, pp. 1064–1065). ACLA displayed "The Deadly Dozen" poster at events, and the poster also appeared in literature published by the Advocates for Life Ministries (ALM), another Portland-based antiabortion group, as well as other individuals opposing abortion.

Following the release of its "Deadly Dozen" poster, ACLA released six other "wanted"-type "guilty" posters in August 1995. Three of these "guilty" posters identified individual doctors, along with their home and work addresses, while the other half identified reproductive health care clinics. Like "The Deadly Dozen" poster, these posters bore antiabortion rhetoric, again including the phrases "ABORTIONIST" and "GUILTY OF CRIMES AGAINST HUMANITY" (*PPCW*, 2002, p. 1065). Also like "The Deadly Dozen" poster, these posters offered a cash reward, in this case to any ACLA organization that could successfully persuade, within the group's guidelines, a doctor to stop performing abortions.

Prior to the release of ACLA's "guilty" posters, several individuals who worked at abortion clinics had been killed after being listed on similar type posters. The murders typically took place at the individuals' places of employment. Fearing similar violence, the plaintiffs in *PPCW*, four physicians and two abortion clinics identified in the antiabortion materials previously discussed, brought a suit under the Freedom of Access to Clinics Entrance Act (FACE). FACE provides that aggrieved persons can take action against any party who uses a "threat of force" to intentionally intimidate any individual because that individual is, or has been, providing reproductive health services. The plaintiffs in *PPCW* alleged that they had been targeted with threats by ACLA, ALM, and numerous individuals (hereinafter antiabortionists).

At trial, a federal district court was faced with the task of construing, in a manner consistent with the First Amendment, what FACE meant by "threat of force," and jurors were asked to determine whether the antiabortionists' speech constituted such a threat. In defining "threat of force," the district court adopted a "true threats" test formulated by the Ninth Circuit of Appeals in *Roy v. United States* (1969). Like *Watts*, *Roy* involved a question of whether a defendant's statement constituted a threat to the president.

In accordance with the *Roy* Court's "true threats" test, the district court determined that under FACE, "threat of force" meant a statement that "a reasonable person" would foresee as being interpreted by those to whom it is communicated "as a serious expression of intent to harm" (*PPCW*, 2002, p. 1062). Additionally, the district court held that, although the antiabortionists' materials did not include any explicit threats, context could be taken into consideration. Operating under these guidelines, a jury found for the plaintiffs, awarding each of them thousands of dollars in compensatory damages and millions in punitive damages. The jury determined that each of the antiabortionists had violated FACE by communicating "true threats" of force in order to intentionally intimidate the plaintiffs from providing legal reproductive health services. Subsequently, the district court barred the antiabortionists from publishing or disseminating the protested materials or providing any other materials that were specifically intended to threaten the plaintiffs. The antiabortionists appealed this ruling to the Ninth Circuit Court of Appeals. The antiabortionists claimed that their materials, which contained no explicitly threatening language, were protected political speech under *Watts*, and could not lose this status because of a context of violence that included the independent action of third parties.

On appeal, a panel of judges from the Ninth Circuit Court of Appeals reversed the lower court's ruling. In the Circuit Court's view, the district court had allowed jurors to find the antiabortionists liable for bringing the plaintiffs to the attention of violent third parties (conduct protected under the Constitution) rather than for the constitutionally proscribable acts of sanctioning or directly threatening harm itself. Thus, the Circuit Court decided that it should independently evaluate the facts of the case to determine whether the content of the antiabortionists' materials could reasonably be interpreted as communicating that the antiabortionists, or their agents, would physically harm physicians who continued to perform abortions. Based on its own analysis of the record, the Circuit Court determined that the antiabortionists' rhetoric had not violated FACE and that the jury's verdict must be overturned.

However, in an unusual move, the Circuit Court later decided that it should rehear *PPCW en banc*, with all of its justices, because of the importance of the issues the case presented. Upon rehearing *PPCW*, the Circuit Court reversed its earlier decision, and affirmed the district court's ruling as to the antiabortionists' guilt, although it remanded the case for consideration of whether the multimillion dollar punitive damages award was consistent with the requirements of due process. In determining that the antiabortionists' materials were not political speech protected by the First Amendment, the Circuit Court also adopted the "true threats" test it had formulated in *Roy*. Following the test, the Circuit Court concluded that, given the previous pattern of antiabortion violence following the release of "wanted"-type posters, it was reasonable to assume that the antiabortionists were aware that their materials would likely be interpreted by the plaintiffs as a "true threat" of death or physical harm. Like the Supreme Court in *Watts* and *Black*, the Circuit Court held that context must be taken into consideration along with content when applying a "true threats" test. As Justice Rymer, writing for the Circuit Court, opined, "context is critical in a true threats case and history can give meaning to the medium" (*PPCW*, 2002, p. 1078). After viewing the challenged materials against the historical backdrop of antiabortion violence that followed the release of similar content, the Circuit Court held that even in the absence of any explicitly threatening language, the antiabortionists' media of expression effectively communicated a "true threat" to the plaintiffs.

As to the "Nuremberg Files" specifically, the Circuit Court felt that the Web site was protected by the First Amendment inasmuch as it contained antiabortion rhetoric. However, the Circuit Court held that the site lost its protected status at the point at which it listed abortion providers and kept track of their physical statuses. As Rymer wrote, "In conjunction with the 'guilty' posters, being listed on a Nuremberg Files scorecard for abortion providers impliedly threatened physicians with being next on a hit list. To this extent only, the Files are also a true threat" (*PPCW*, 2002, p. 1088). Thus, just as the Supreme Court would later articulate in *Black*, the Circuit Court held that "true threats" include not only serious expressions of an intention to perpetrate illegal acts of violence, but also speech made with intent merely to intimidate. The Supreme Court declined to review this case (*American Coalition of Life Activists v. Planned Parenthood of the Columbia/Williamette*, 2003).

Of all the cases discussed thus far, *PPCW* is the most relevant to this study. Not only did *PPCW* establish that hateful speech posted in publicly accessible areas online can be constitutionally regulated, it also established that a "true threats" test is applicable to hate sites. If a court can use a "true threats" test to evaluate the constitutionality of hateful rhetoric contained on an antiabortion site, certainly it can use the same standard to assess the lawfulness of content appearing on a Web site maintained by a hate group with, say, an anti-Semitic sentiment. Additionally, *PPCW* set the precedent that threatening Internet rhetoric can be proscribed, even when the language is merely implicit and not directly linked to illegal action. This may have important First Amendment implications for hate sites, given hate groups' propensity for using threatening language, as evidenced in the *Brandenburg* case. Thus, *PPCW* is particularly relevant to the issue of regulating hate sites.

The cases of *Baker* and *PPCW* leave little doubt that a "true threats" test is applicable to hateful online communications, whether delivered in private e-mails or posted on publicly accessible Web sites. Indeed, there are some legal scholars who believe that a "true threats" test is the most appropriate standard for analyzing the constitutionality of Internet hate speech (e.g., Brenner, 2002). However, a "true threats" test is not the only First Amendment standard adopted by a federal court in an Internet speech case that can be used to evaluate the constitutional validity of hate sites. In the interim between the *Baker* and *PPCW* decisions, the Supreme Court, in its first ruling relating to online speech, adopted a First Amendment test formulated in a pre-Internet case that could prove extremely useful in determining whether a given hate site is constitutionally protected.

**The "*Miller* Test"**

While several federal courts had already made rulings in the area of online speech, the Supreme Court did not weigh in on an Internet-related case until 1997 when it handed down its landmark decision in *Reno v. ACLU* (1997). At issue in *Reno* was the constitutionality of the "indecent transmission" and "patently offensive display" provisions of the Communications Decency Act of 1996 (CDA), Title V of the seven-part Telecommunications Act of 1996. The "indecent transmission" provision of the CDA, Section 223(a), provided:

> Whoever...(1)...(B) by means of a telecommunications device knowingly...(i) makes, creates, or solicits, and...(ii) initiates the transmission of,...any comment, request, suggestion, proposal, image, or other communication which is obscene or indecent, knowing that the recipient of the communication is under 18 years of age, regardless of whether the maker of such communication placed the call or initiated the communication; [or]...(2) knowingly permits any telecommunication facility under his control to be used for any activity prohibited by paragraph (1) with the intent that it be used for such activity...shall be fined...or imprisoned not more than two years, or both. (*Reno*, 1997, p. 859)

Similarly, the CDA's "patently offensive display" provision, Section 223(d), stated:

> Whoever...(1)...(A) uses an interactive computer service to send to a specific person or persons under 18 years of age, or...(B) uses any interactive computer service to display in a manner available to a person under 18 years of age, ...any comment, request, suggestion, proposal, image, or other communication that, in context, depicts or describes, in terms patently offensive as measured by contemporary community standards, sexual or excretory activities or organs, regardless of whether the user of such service placed the call or initiated the communication; or...(2) knowingly permits any telecommunication facility under such person's control to be used for an activity prohibited by paragraph (1) with the intent that it be used for such activity...shall be fined...or imprisoned not more than two years, or both. (*Reno*, 1997, p. 860)

By replacing the term "telephone" with "telecommunications device" and "interactive computer service," Sections 223(a) and 223(d) amended protections that existed in the Communications Act of 1934 such that online communication could also be regulated (Meadows, 1996).

Immediately after the CDA went into effect, the ACLU and 19 other plaintiffs filed a suit challenging the constitutionality of sections 223(a)(1) and 223(d). Shortly thereafter, 27 other plaintiffs filed a similar suit. The U.S. District Court for the Eastern District of Pennsylvania consolidated the two cases. In a unanimous decision, the three-judge court held that sections 223(a)(1) and 223(d) would abridge a significant amount of protected speech and barred the government from enforcing any aspect of the provisions, except for 223(a)'s ban on obscenity. The reasoning that the District Court provided for its decision would be an important factor later on when the case went before the Supreme Court.

On direct appeal, the Supreme Court affirmed the District Court's ruling. The Court held that the CDA's "indecent transmission" and "patently offensive display" provisions were unconstitutionally vague and restricted an unacceptably high level of expression protected by the First Amendment. Justice Stevens, who delivered the Court's opinion, stated that the ambiguous language of the provisions, coupled with their threat of severe criminal sanctions, would undoubtedly cause some Internet speakers to refrain from engaging in constitutionally protected expression. In arriving at its decision, the Court, instructed by the findings of the Pennsylvania District Court, made several determinations that established the ground rules for regulating the Internet in the United States.

For one, the Court established that the Internet could not be regulated like radio or television. In earlier cases, the Supreme Court had determined that broadcast media were only entitled to a limited level of First Amendment protection (e.g., broadcast-specific regulations) since they used frequencies which were scarce in number. What is more, the Supreme Court had held that broadcast media should only be given a limited degree of First Amendment protection because their signals were invasive, able to freely enter into any area equipped with an antenna. The *Reno* Court recognized, however, that the "scarcity" and "invasiveness" rationales that the Supreme Court had used to justify the regulation of broadcast media did not apply to a non-broadcast medium like the Internet. With respect to the "scarcity" rationale for content regulation, Stevens wrote, "the Internet can hardly be considered a 'scarce' expressive commodity. It provides relatively unlimited, low-cost capacity for communication of all kinds" (*Reno v. ACLU*, 1997, p. 870). Regarding the "invasiveness" rationale for regulating speech, Stevens stated, "the Internet is not as 'invasive' as radio or television" (*Reno*, 1997, p. 869). Quoting from the District Court's opinion, Stevens explained, "communications over the Internet do not 'invade' an individual's home or appear on one's computer screen unbidden. Users seldom encounter content 'by accident'" (*Reno*, 1997, p. 869). Thus, much to the delight of advocates of online speech, the Court recognized that the Internet was a unique medium that demanded regulation tailored to its uniqueness.

Additionally, the Court held that any statute aimed at protecting minors from harmful Internet content must conform to the established legal precedents in the area of indecent and obscene speech regulation. In several previous cases, the Court had held that, although government has a compelling interest in shielding children from harmful materials, statutes regulating indecency must be narrowly drafted to avoid unnecessarily restricting adults' constitutional right to access and distribute indecent communications. Stevens indicated that this same logic applied to the Internet when he wrote, "In order to deny minors access to potentially harmful speech, the CDA effectively suppresses a large amount of speech that adults have a constitutional right to receive and to address to one another… . the CDA is not narrowly tailored" (*Reno v. ACLU*, 1997, pp. 874, 879).

In determining that the CDA was not narrowly tailored, the Court specifically rejected the government's claim that the age verification defenses provided by Section 223(e)(5) of the Act would significantly reduce any heavy burden placed on adult speech by its "indecent transmission" and "patently offensive display" provisions. Section 223(e)(5) stated in part:

> It is a defense to a prosecution…that a person…(A) has taken, in good faith, reasonable, effective, and appropriate actions under the circumstances to restrict or prevent access by minors to a communication specified in…[Sections 223(a) and 223(d)]… which may involve any appropriate measures to restrict minors from such communications, including any method which is feasible under available technology; or…(B) has restricted access to such communication by requiring use of a verified credit card, debit account, adult access code, or adult personal identification number. (*Reno v. ACLU*, 1997, p. 861)

The Court held, however, that these defenses were not enough to save the CDA's otherwise unconstitutional "indecent transmission" and "patently offensive display" provisions.

To begin with, the Court pointed out that there was virtually no reliable way to verify the age of persons accessing content in most Internet forums, including sites posted on the Internet. Hence, adults fearing prosecution under the nonobscenity provisions of the CDA would still have been effectively barred from the online distribution of indecent content that they have a constitutional right to address to one another offline. Additionally, the Court held that while it was technologically feasible for online content providers to require users to provide information like an access code to gain access to their Web sites, the costs of maintaining such a verification system would likely force many noncommercial Web sites to shut down. Consequently, as Stevens suggested, Internet content providers lacking a reliable or practical means of verifying users' ages may have refrained from disseminating materials that could have been deemed "indecent" or "patently offensive," thus reducing online discourse to a level suitable only for children. The Court also reasoned that adults who did not have a credit card and lacked the resources to obtain one would have been completely prohibited from accessing any indecent online material that was blocked. Even where a credit or debit card was not required, the Court held that an adult password requirement (i.e., access code, personal identification number) may have discouraged adults from accessing indecent material online, especially from noncommercial sites. In essence, the Court found that, while the interest in protecting children from harmful materials is compelling, in the absence of a viable process for verifying Internet users' ages, it is unconstitutional to prohibit adults from transmitting or receiving indecent content online. Several other federal courts that have assessed the constitutionality of efforts at regulating Internet speech have reiterated this principle.

Although Stevens indicated that Internet indecency can only be regulated to a limited degree, he made it clear that an absolute prohibition against obscene online content, including child pornography, would be permissible since obscenity is unconstitutional, a conclusion that the Supreme Court first arrived at in *Roth v. United States* (1957). However, Stevens suggested that any statute aimed at prohibiting the exchange of obscene materials on the Internet must utilize the test for defining obscenity that the Court developed 24 years earlier in the case of *Miller v. California* (1973).

*Miller* involved a commercial vendor's conviction for mailing unsolicited advertising brochures for "adult" materials in violation of a California statute that made it a misdemeanor to knowingly distribute "obscene" matter. In overturning the defendant's conviction, the Court, hoping to provide a more precise definition of "obscene" than it had in *Roth*, formulated a three-prong test for defining obscenity, known today simply as the "*Miller* test." Under the *Miller* test, portions of which were adopted from *Roth*, for materials to be considered "obscene" depended on:

> (a) whether the "average person, applying contemporary community standards" would find that the work, taken as a whole, appeals to the prurient interest...; (b) whether the work depicts or describes, in a patently offensive way, sexual conduct specifically defined by the applicable state law; and (c) whether the work, taken as a whole, lacks serious literary, artistic, political, or scientific value. (*Miller*, 1973, p. 24)

Although the CDA contained some of the same language (e.g., "patently offensive," "contemporary community standards") found in this three-prong test, Stevens explained that the *Reno* Court believed that the statute, as a whole, provided a much broader definition of obscenity than that set forth in *Miller*. However, since obscenity was nevertheless devoid of constitutional protection, the Court severed the phrase "or indecent" from Section 223(a) of the CDA, leaving the rest of the provision standing as it related to obscene content. Stevens pointed out that this was the only way that the CDA's constitutionality could be saved.

While Stevens discussed the *Miller* test in *Reno*, some First Amendment scholars claimed that the Court never adopted the standard for regulating online communications because it recognized that the test's "contemporary community standards" provision could not be applied to the Internet, a medium without boundaries. However, statements made by the Court in an Internet speech case 5 years later belied this contention.

In response to the *Reno* Court's decision, Congress passed the Child Online Protection Act (COPA), which prohibited online content providers from knowingly making available to a minor any material that they know to be "harmful to minors." Just as the CDA imposed criminal sanctions, COPA provided that persons violating the Act could be fined up to $50,000 or sentenced to as much as 6 months in prison, or both. Also like the CDA, COPA provided an affirmative defense to persons subject to prosecution under the Act. In language reminiscent of the CDA's age verification defense, COPA's affirmative defense stated that a person could avoid conviction under the Act by demonstrating that he or she:

> [H]as restricted access by minors to material that is harmful to minors...(A) by requiring the use of a credit card, debit account, adult access code, or adult personal identification number; (B) by accepting a digital certificate that verifies age; or (C) by any other reasonable measures that are feasible under available technology. (*Ashcroft v. ACLU*, 2002, p. 570)

Despite its many similarities to the CDA, COPA also had many important differences.

While the CDA covered all online communications, COPA applied only to material displayed for "commercial purposes" on the Internet. Additionally, whereas the CDA regarded minors as persons younger than 18 years of age, COPA defined a "minor" as an individual younger than 17. COPA was also unique in that it made individuals subject to civil sanctions of up to $50,000 for each violation of the Act. However, the most important way in which COPA differed from the CDA was in how it defined the categories of speech that were prohibited under the Act. In drafting COPA, Congress followed Stevens' suggestion in *Reno* and drew upon the *Miller* test for obscenity. With some modifications to the language of the *Miller* test, COPA defined content "harmful to minors" as:

> [A]ny communication, picture, image, graphic image file, article, recording, writing, or other matter of any kind that is obscene or that...(A) the average person, applying contemporary community standards, would find, taking the material as a whole and with respect to minors, is designed to appeal to, or is designed to pander to, the prurient interest;...(B) depicts,

> describes, or represents, in a manner patently offensive with respect to minors, an actual or simulated sexual act or sexual contact, an actual or simulated normal or perverted sexual act, or a lewd exhibition of the genitals or post-pubescent female breast; and...(C) taken as a whole, lacks serious literary, artistic, political, or scientific value for minors. (*Ashcroft v. ACLU*, 2002, p. 570)

Like the CDA, COPA was written broadly enough (i.e., "any communication") to encompass obscene communications over the Internet. However, unlike the CDA, COPA was narrowly tailored to avoid unnecessarily burdening adult speech online. While still prohibiting the unprotected category of obscenity, COPA's other restriction applied not to "indecent" and "patently offensive" material, but to the narrower class of content considered harmful only "with respect to minors" age 17 or younger. Although Congress seemingly gave much consideration to the *Reno* Court's holding when drafting the Act, COPA produced even more litigation than did the CDA.

Before the Act had even gone into effect, the ACLU and 17 other plaintiffs filed a lawsuit challenging COPA's constitutionality in the U.S. District Court for the Eastern District of Pennsylvania. The plaintiffs alleged that they feared being prosecuted under the Act because some of the content on their Web sites, which they believed was valuable for adults, might be construed as being "harmful to minors" by certain communities' standards. Therefore, the plaintiffs claimed that COPA violated adults' First and Fifth Amendment rights because it: (1) effectively banned speech that adults have a constitutional right to send and receive, (2) was not narrowly tailored to serve any compelling governmental interest, and (3) was significantly overbroad.

As it had done before when asked to assess the constitutionality of the CDA, the Pennsylvania District Court ruled in favor of the ACLU and its fellow plaintiffs and entered a preliminary injunction against enforcement of COPA. The government appealed this ruling to the United States Court of Appeals for the Third Circuit. However, the Third Circuit Court of Appeals affirmed the District

Court's ruling, albeit for a different reason. Completely ignoring the lower court's holding, the Circuit Court simply held that COPA was rendered unconstitutionally overbroad by its reliance on "community standards" as the basis for determining that content was "harmful to minors." In making its ruling, the Circuit Court took into consideration that Web site access cannot be restricted on a geographic basis. Thus, the Circuit Court reasoned that COPA's application of "community standards" would require Internet publishers to place vast amounts of content behind adult verification systems for fear that extremely conservative communities might deem the material harmful to minors. In short, the Circuit Court concluded that the "community standards" criterion had no applicability to the Internet, a view shared by various academic and legal scholars.

After granting the government's request to review the Circuit Court's decision regarding the constitutionality of COPA, the Supreme Court reversed in the case of *Ashcroft v. ACLU* (2002) (hereinafter *Ashcroft I*). In an opinion authored by Justice Thomas, the high court held that, even though applied to the nationwide communication medium of the Internet, COPA's use of the "community standards" criterion did not render the Act unconstitutionally overbroad. Thomas stated that there was no requirement that a statute regulating speech define "community standards" in precise geographic terms. Therefore, Thomas pointed out that COPA could not be declared unconstitutional simply because it relied on the "community standards" criterion to regulate content over the Internet, a medium without fixed geographic boundaries. Thomas explained that the Court believed that absent geographic specification, jurors applying "community standards" would inevitably draw upon their personal knowledge of the communities from which they came.

As indicated by Thomas, the Court's holding was consistent with its decisions in two other federal antiobscenity statute cases, *Hamling v. United States* (1974) and *Sable Communications of California v. FCC* (1989). In *Hamling* and *Sable*, the Court upheld the application of the "community standards" criterion to the national communication networks of the postal service and the telephone, respectively.

Relying on these decisions, Thomas responded to the Circuit Court's claim that COPA's reliance on "community standards" would make Internet publishers responsible for abiding by the standards of all the communities into which their content enters. He opined:

> If a publisher chooses to send its materials into a particular community, this Court's jurisprudence teaches that it is the publisher's responsibility to abide by that community's standards. The publisher's burden does not change simply because it decides to distribute its materials to every community in the Nation… . Nor does it change because the publisher may wish to speak only to those in a "community where avant garde culture is the norm,"…but nonetheless utilizes a medium that transmits its speech from coast to coast. If a publisher wishes for its materials to be judged only by the standards of particular communities, then it need only take the simple step of utilizing a medium that enables it to target the release of its materials into those communities. (*Ashcroft v. ACLU*, 2002, p. 583)

Thus, the Court definitively held that, like other exercises of speech, a person's choice to transmit content over the Internet carried with it a responsibility to abide by the standards of each community confronted with his or her communication. So, while *Reno* established that the Internet could not be regulated like radio or television, the *Ashcroft I* Court indicated that the medium could be regulated like the postal and telephone systems inasmuch as a statute restricting online content "harmful to minors" relied on the "community standards" criterion.

The *Ashcroft I* Court was only asked to determine whether COPA's use of "community standards" violated the Constitution. However, Thomas also took the opportunity to explain why the Court believed the Act did not suffer from the same constitutional problems that it found with the CDA. As Thomas explained, key in the Court's determination was the fact that, unlike the CDA, COPA provided a definition of content "harmful to minors" that was adapted from the language of the *Miller* test. He wrote, "COPA… applies to significantly less material than did the CDA and defines the harmful-to-minors materials restricted by the statute in a manner parallel to the *Miller* definition of obscenity" (*Ashcroft I*, 2002, p. 578). Thomas' approval of the "harmful-to-minors materials" definition provided by COPA, coupled with his statements specifically regarding the constitutionality of the Act's use of "community standards," made it clear that the Court believed that the *Miller* test, as a whole, has applicability in the Internet context.

While the *Ashcroft I* ruling revealed that members of the Court supported COPA's use of language adopted from the *Miller* test, it did not settle the issue of the Act's constitutionality. After making its ruling, the *Ashcroft I* Court remanded the case to the Third Circuit Court of Appeals to reconsider whether the Pennsylvania District Court abused its discretion in entering a preliminary injunction against enforcement of COPA. On remand, the Circuit Court concluded that there was no abuse of discretion, and reaffirmed the District Court's earlier holding. This time, however, the Circuit Court did not ignore the grounds on which the District Court based its decision to prohibit the government from enforcing COPA. After considering the District Court's findings, the Circuit Court held that COPA was overbroad and was not narrowly tailored to serve any compelling government interest. What is more, the Circuit Court held that the Act was not the least-restrictive means available to government for serving its interest in protecting minors from online content that was harmful to them. Once again, the Supreme Court granted the government's request to review the Circuit Court's decision with respect to COPA's constitutionality.

However, upon its second review of COPA's constitutionality in *Ashcroft v. ACLU* (2004) (hereinafter *Ashcroft II*), the Supreme Court affirmed the Circuit Court's decision to uphold the preliminary injunction against enforcement of the Act. Whereas the *Ashcroft I* Court's review was limited to the question of whether the use of "community standards" rendered COPA unconstitutional, the Court in *Ashcroft II* assessed the Act's constitutionality in its entirety. The Court considered how COPA defined terms (e.g., "minors," "harmful to minors"), the sanctions the Act imposed, and the affirmative defense provided by the statute. After assessing these various elements of COPA, the Court concluded that the Act did not represent the least-restrictive means of accomplishing the government's compelling interest in protecting minors from harmful online content. As in *Reno*, the Court did not find that COPA's affirmative defense sufficiently reduced the burden the Act placed on the speech rights of adults and Internet publishers.

The existence of Internet filtering technology weighed heavily in the *Ashcroft II* Court's decision. As Justice Kennedy, who delivered the Court's opinion, wrote:

> Filters are less restrictive than COPA. They impose selective restrictions on speech at the receiving end, not universal restrictions at the source. Under a filtering regime, adults without children may gain access to speech they have a right to see without having to identify themselves or provide their credit card information. Even adults with children may obtain access to the same speech on the same terms simply by turning off the filter on their home computers. Above all, promoting the use of filters does not condemn as criminal any category of speech, and so the potential chilling effect is eliminated, or at least much diminished. (*Ashcroft II*, 2004, p. 2792)

Essentially, the Court reasoned that it would be a less-restrictive burden on the speech rights of Internet publishers and adult Internet users to simply make adults responsible for filtering out online content that they do not want entering their homes.

While the Court upheld the preliminary injunction against COPA's enforcement, it did not rule the Act unconstitutional per se. Rather, the Court remanded the case to the Pennsylvania District Court for a trial in which the government would be given the opportunity to show that COPA is the least-restrictive means available to Congress for protecting minors from harmful online content. This move prompted Justice Breyer, writing for Justices Rehnquist and O'Connor in his dissenting opinion, to ask: "What remains to be litigated?" (*Ashcroft II*, 2004, p. 2805). Well, apparently something remained to be litigated. At the time this book was being written, the Pennsylvania District

Court was once again asked to rule on COPA's constitutionality in the case of *ACLU v. Gonzales* (2007). Not surprisingly, the District Court once again ruled in favor of the ACLU and its fellow plaintiffs and entered an injunction against enforcement of the Act. The ruling will likely be appealed to the U.S. Supreme Court, giving the high court its third opportunity to examine issues related to COPA. If the high court agrees to hear an appeal of *ACLU v. Gonzales*, perhaps it will finally rule COPA unconstitutional and put a stop to the never-ending litigation involving the Act.

Whatever the ultimate fate of COPA may be, it seems clear that the *Ashcroft II* decision reaffirmed, albeit implicitly, that the Supreme Court still believes the *Miller* test is applicable to the Internet. Although the Court could have ended all future COPA litigation simply by declaring that its *Miller*-test-like "harmful to minors" provision rendered the Act unconstitutional, the majority declined to do so. By refusing to do so, the *Ashcroft II* Court reestablished that the *Miller* test can be used to assess the constitutionality of online content.

Still, some argue that the *Miller* test has no applicability to the Internet given its reliance on "contemporary community standards" as a basis for regulating speech. However, the *Reno*, *Ashcroft I*, and *Ashcroft II* decisions demonstrate that the Supreme Court has taken the opposite view. Time and time again, the justices of the Supreme Court have indicated their support for the application of the *Miller* test in the Internet context. It seems evident that the *Miller* standard, while established well before the advent of the Internet, is a First Amendment test that has been adopted by the high court for regulating online content. And while it was not designed for assessing hateful speech, the *Miller* test can still prove useful in evaluating the constitutionality of hate sites (Leets, 2001). Indeed, Leets relied on questions adapted from the *Miller* test in her study of the constitutionality of hate sites maintained by white supremacist groups. Leets' reliance on questions adapted from the *Miller* test is well-founded. Because while the hate speech debate has centered around whether this type of speech causes harm sufficient to carry it outside of First Amendment protection, if there is obscene material on hate sites, that alone would be enough to render them unconstitutional.

## Summary

Although the Internet is a unique medium of communication, several federal courts have determined that some of the First Amendment tests developed in pre-Internet cases are also applicable to online content. There are two First Amendment tests that have been adopted by the federal courts for assessing Internet content that were relevant to this study: a "true threats" test and the *Miller* test for obscenity.

Five years after the commercialization of the Internet, it became evident that the online world was not immune from hateful speech. Required to assess the constitutionality of hateful speech about women delivered via private e-mail, a federal district court in *United States v. Baker* elected to adopt a "true threats" test developed in a pre-Internet case by the Second Circuit Court of Appeals. In the wake of *Baker*, other federal courts have also adopted "true threats" tests to evaluate hateful materials transmitted via private e-mails and posted on publicly accessible Web sites, the focus of this study. It is clear that, regardless of what court drafted the standard, a "true threats" test has been adopted by the federal courts for use in analyzing allegedly threatening online content.

Two years after *Baker*, the Supreme Court would consider the issue of Internet speech in *Reno v. ACLU*, a case in which it assessed the constitutionality of the CDA's "indecent transmission" and "patently offensive display" provisions. While the CDA was struck down, of importance to this study is that, in assessing the constitutionality of the Act's challenged provisions, the Court adopted the *Miller* test for obscenity, yet another First Amendment standard developed in a pre-Internet case. Since *Reno*, the Court has repeatedly indicated that it believes the *Miller* test is applicable to Internet content.

The "true threats" and *Miller* tests are extremely important standards to use in assessing the constitutionality of hate sites. To begin with, both tests serve as Internet-specific First Amendment standards for evaluating hate sites. Secondly, as the tests were designed to apply to different categories of speech, they allow the researcher to subject significant amounts of hate site content to First Amendment analysis. While a "true threats" test can be utilized to assess whether there is nonobscene threatening content on hate Web sites, the *Miller* test can be used to determine whether there is nonthreatening obscene materials on the same sites.

# Chapter 6

# Applicability of Standards for Regulating Hateful Speech in Non-Internet Settings to Hate Speech Posted Online

The Internet has called into question many of the doctrines that define modern free speech jurisprudence (Gey, 1999). While the *PPCW* Court basically established that a "true threats" test is pertinent in evaluating hate sites, the question remains whether the other First Amendment standards applied to hateful speech communicated in non-Internet settings are also applicable to hate site content. This chapter discusses the extent to which the First Amendment tests reviewed in Chapter Four are applicable to hate speech posted on the Internet. This analysis was informed by the commentaries of legal scholars and the rationales for proscribing speech provided by federal courts in precedent-setting freedom of expression cases, especially those in which the First Amendment tests were originally developed. The chapter begins by reviewing those tests that analysis demonstrates are not applicable to hate sites and then turns to a discussion about the tests that analysis reveals can be applied to hate speech posted on the Internet.

## FIRST AMENDMENT TESTS FOR HATEFUL SPEECH THAT ARE INAPPLICABLE TO HATE SITES

### "Fighting Words" Test

There are those who claim that some kinds of hate speech (e.g., racist speech directed toward persons or small groups of persons) qualify as unprotected "fighting words." Regardless of whether this is true, the "fighting words" test is clearly not applicable to hate sites or any other content disseminated via the Internet. Even if a statute could be drawn to satisfy all the elements necessary to sustain a conviction under the test, *Chaplinsky's* threshold requirement that the words be voiced during a *face-to-face* encounter would obviously not be satisfied by any material appearing online.

Still, the First Amendment commentary provided by the Supreme Court in its "fighting words" cases is relevant to this study. For example, Tsesis (2001) points out, "Any cause of action against hate speech on the Internet must address the seminal case on this point, *R.A.V. v. St. Paul*" (p. 849). Therefore, *R.A.V.* and the other "fighting words" cases still informed the analysis regarding the constitutionality of the hate sites in this study.

### "Group Defamation" Test

Like the "fighting words" test, the "group defamation" standard is among the most commonly advanced justifications for regulating hate speech, despite the fact that the standard's usefulness as a legal precedent is tenuous at best. It is true that *Beauharnais*, the case in which the "group defamation" test was formulated, remains good law in theory. However, many legal scholars regard *Beauharnais* as irrelevant (O'Callaghan, 1994). As Massey (1992) writes, "*Beauharnais v. Illinois*, the only extant constitutional doctrinal foundation for suppression of group libel, has been virtually destroyed as useful precedent" (p. 166). Massey's assessment of *Beauharnais'* validity is representative of a broad spectrum of legal opinion. Some who consider *Beauharnais* irrelevant note that the Supreme Court decided the case prior to its decisions in *Brandenburg*, which required that speech be evaluated on the basis of its likelihood to produce imminent lawlessness, and *New York Times v. Sullivan* (Barron & Dienes, 2004).

*New York Times v. Sullivan* (1964) is the landmark ruling in the area of libel law. The case involved a half-million dollar civil libel action brought by L. B. Sullivan, one of Montgomery, Alabama's three elected city commissioners, against the *New York Times* and four black clergymen from Alabama. Sullivan, whose duties included supervising the Montgomery police department, claimed that he had been defamed by a full-page advertisement that ran in the *Times* in March 1960. The advertisement, which was a call for financial support for the civil rights movement in the South, included some erroneous statements about Montgomery police handling of particular civil rights activities and activists. Also appearing in the advertisement were the names of 20 persons, including those of the 4 individual defendants, who were identified as endorsing the appeal for financial support contained in the ad. Although the statements at issue in the case never mentioned him by name, Sullivan claimed that the false charges about police conduct libeled him personally since they could be read as referring to his performance as Montgomery police supervisor. An Alabama jury agreed with Sullivan, and awarded him the $500,000 in damages he had sought, a determination which Alabama's Supreme Court would later affirm.

However, on appeal, the U.S. Supreme Court unanimously reversed. The Court held that public officials could only recover damages for defamatory falsehoods relating to their official conduct if they could prove that the statements were made with "actual malice," which it defined as having knowledge that statements were false or having reckless disregard for whether they were false. In making this determination, the *Sullivan* Court significantly reduced the field of libelous speech that can be constitutionally proscribed. Much of what citizens publicly criticize is the conduct of public officials, and this criticism often includes factual inaccuracies, as did the advertisement at issue in *Sullivan*. As Justice Brennan, speaking for the Court, wrote, "erroneous statement is inevitable in free debate, and…it must be protected if the freedoms of expression are to have the 'breathing space' that they 'need…to survive' " (*New York Times v. Sullivan*, 1964, pp. 271–272).

Thus, the *Sullivan* Court extended substantial First Amendment protection to libelous statements (Barron & Dienes, 2004), protection which should also extend to the area of group libel. Surely, if the Court recognizes that erroneous statement is inevitable in a discussion about an individual then it must recognize the inevitability of falsehoods being stated during discussions about a group of individuals, especially one as large as was being referred to in the literature at issue in *Beauharnais*. Consequently, *Sullivan* considerably decreased the force of the libel prong of the three-part "group defamation" test that was used in upholding the defendant's conviction for group libel in *Beauharnais*, and, by extension, essentially rendered the case all but useless as a legal precedent. *Beauharnais'* usefulness as precedent was even further diminished by the fact that, following *Sullivan*, the Court extended the "actual malice" requirement to "public figures," individuals who often serve an important function in ordering society even though they are not governmental officials (Cate, 1995).

Another case that signified the demise of *Beauharnais* is *Collin v. Smith* (1978). The case involved the efforts of the National Socialist Party of America (NSPA), a Nazi group, to march in the village of Skokie, Illinois, a Chicago suburb with a large Jewish population, including several thousand Holocaust survivors. In May 1977 NSPA and its leader, Frank Collin, announced that the party planned to march in front of Skokie's village hall, an announcement that sparked a firestorm of litigation. Skokie officials responded by getting a state trial court to issue a preliminary injunction against the march. After a state appeals court and the Illinois Supreme Court both denied them an application for a stay of the injunction, NSPA filed an application with the U.S. Supreme Court (*National Socialist Party of America v. Village of Skokie*, 1977). In a brief *per curiam* opinion, the Court held that NSPA would be deprived of its First Amendment rights during a period of appellate review, and granted the group's application for a stay of the injunction and remanded the case for further proceedings. As a result of the Court's ruling, the injunction was eventually reversed (*Village of Skokie v. National Socialist Party of America*, 1978).

However, while all this litigation was taking place Skokie had enacted three statutes to prohibit demonstrations of the sort Collin and NSPA had threatened. One of the statutes, No. 77-5-N-994 (hereinafter 994), which had to do with insurance requirements, indicated that Skokie officials would only grant a permit for a march if they found that the proposed demonstration would not: "portray criminality, depravity or lack of virtue in, or incite violence, hatred, abuse or hostility toward a person or group of persons by reason of reference to religious, racial, ethnic, national or regional affiliation" (*Collin v. Smith*, 1978, p. 1199). The ordinance's language was adapted from the Illinois group libel law at issue in *Beauharnais*. Another statute, No. 77-5-N-995 (hereinafter 995), similarly prohibited persons from disseminating within Skokie any materials which were intended to promote and incite hatred against individuals on the basis of their race, religion, or national origin. As defined by the statute, "dissemination of materials" included publicly distributing literature or displaying materials, markings, or clothing of symbolic significance. The final ordinance, No. 77-5-N-996 (hereinafter 996), prohibited "public demonstrations by members of political parties while wearing 'military-style' uniforms" (*Collin v. Smith*, 1978, p. 1200).

Undeterred by Skokie's enactment of its three demonstration ordinances, Collin and NSPA twice applied for permission to march in the village on July 4, 1977. NSPA's first permit application stated that its march would involve 30 to 50 individuals wearing the group's Nazi-style military uniforms and carrying the party's banner. NSPA indicated that its banner would include a swastika and placards with comments like "White Free Speech," "Free Speech for White America," and "Free Speech for the White Man" (*Collin v. Smith*, 1978, p. 1200). NSPA further proposed that it would march single file on the sidewalk, and would refrain from giving speeches or distributing literature, so as not to disrupt traffic in Skokie. NSPA's second permit application contained similar recitations.

Skokie did not maintain that NSPA would behave other than as was described in its permit applications. Nevertheless, Skokie denied NSPA's applications to march because the group supposedly disclosed an intention to violate ordinance 996, and the litigation commenced once again.

Collin and NSPA filed a lawsuit in a federal district court in an attempt to prohibit Skokie and its various officials, including village President Albert Smith, from enforcing the village's various demonstration ordinances. The district court granted relief to Collin and NSPA, holding that 994's insurance requirements were an invalid restriction on free speech, that 995 was impermissibly vague and overbroad, and that 996 was unconstitutionally overbroad and patently unjustified. Skokie then appealed this decision to the United States Court of Appeals for the Seventh Circuit. In turn, the Seventh Circuit Court, in the case of *Collin v. Smith* (1978), affirmed the lower court's ruling with regard to the unconstitutionality of Skokie's demonstration ordinances. The Court held that the ordinances violated the constitution by prohibiting the peaceful expression of views that, while unpopular, were nevertheless protected by the First Amendment.

Whatever might have been the situation before, the Seventh Circuit's ruling, coupled with the Supreme Court's stay of the original injunction barring NSPA's march, indicated that federal courts did not place much reliance on *Beauharnais*. After all, the Circuit Court declared unconstitutional an ordinance that was adapted from the law upheld in *Beauharnais*. What is more, the Circuit Court implicitly rejected one of the main theories the *Beauharnais* Court offered for upholding group libel laws, the danger that hate speech directed at a defined group could produce intergroup violence (Barron & Dienes, 2004). Clearly, NSPA's intent in planning to march through Skokie in Nazi-style military uniforms, while displaying a banner bearing a swastika, was to direct a form of symbolic anti-Semitic speech at the village's predominantly Jewish population that would have a powerful emotional impact on them. Furthermore, given the Nazis' active role in the mass genocide of Jews during the Holocaust, NSPA's pro-Nazi speech could certainly have affected Skokie's Jewish residents so strongly that they may have been stirred to violence against the group. In fact, Skokie's Jewish residents may have even been stirred to violence against other non-Jewish groups in the area. Yet, even though it appears that NSPA's march could have been constitutionally banned under the "danger theory" that Frankfurter invoked in *Beauharnais*, the Seventh Circuit Court did not rely on the case as precedent. Indeed, the Court stated that, after *Sullivan* and the Supreme Court's other constitutional libel cases, it shared other federal court's doubts about whether "*Beauharnais* remains good law at all..." (*Collin v. Smith*, 1978, p. 1205).

Ironically enough, after finally winning its protracted legal battle, NSPA never marched in Skokie (Fogo-Schensul, 1997/1998). Nevertheless, NSPA's legal victory established a precedent that other hate groups continue to benefit from, as evidenced by the ongoing public demonstrations given by hatemongers on America's streets, sidewalks, and courthouse grounds.

In view of the precedents established by the decisions in *Sullivan* and *Smith*, it is hard to imagine that *Beauharnais* would be decided the same way today (Spector, 2002). However, even if *Beauharnais* remains good law, the three-part "group defamation" test that grew out of the case is still not applicable to the Internet by reason of its "danger" prong. In formulating the "danger" prong of the test in *Beauharnais*, Frankfurter looked at the incidents of speech-induced intergroup violence in Illinois preceding the defendants' prosecution for distributing racist literature that included a call for the white residents of Chicago to unite in opposing black Chicagoans. In light of that history, Frankfurter reasoned that Illinois was justified in attempting to restrict the public communication of defamatory, and potentially inflammatory, statements about "racial and religious groups" (*Beauharnais*, 1952, p. 261). Hence, an examination of *Beauharnais* reveals that the "group defamation" test's danger prong was intended to apply only to speech thought likely to generate group violence. This makes the "group defamation" test's application to the Internet totally inappropriate as there is virtually no likelihood of group violence ever resulting from materials being posted online. As Siegel (1999) writes, "enforcement of group libel statutes over the Internet is questionable" (p. 387). A requirement that speech be likely to produce group violence renders yet another one of the First Amendment tests for regulating hateful speech in non-Internet settings inapplicable to online communications.

**"Imminent Lawless Action" Test**

Although many legal scholars consider it to be the current First Amendment standard for regulating hate speech, *Brandenburg*'s "imminent lawless action" test is of dubious application to hate sites (Brenner, 2002; Leets, 2001). Like the "group defamation" standard, the "imminent lawless action" test was only intended to be utilized for assessing communications believed likely to create group violence, which again is a virtual impossibility with Internet content. As Schlosberg (2001) writes:

> [I]t is difficult to imagine that communications on the Web could meet the [*Brandenburg*] test of illegality. The test, originally created to allow legislators to criminalize speech that would incite a riot, does not seem applicable to cyberspace where riots are nearly impossible. (p. 75)

Indeed, the Supreme Court has not always found that hostile, and arguably inflammatory, statements made during large-scale demonstrations in offline settings, where riots are very possible, meet the test of illegality established in *Brandenburg*. *Hess v. Indiana* (1973) provides a good example. In *Hess*, the Court failed to find that loud utterances made during a public antiwar demonstration, which indicated a desire for protesters to reclaim a street that was being taken over by authorities, were likely to incite a group of more than 100 demonstrators to riot. It is thus difficult to imagine how such a statement, or any other content that could be interpreted as advocating that others engage in acts of lawlessness, being posted on the Internet could be found sufficient to incite group violence.

Even if the "imminent lawless action" test's application was not limited to speech believed liable to induce group violence, the standard is still inapplicable to hate sites due to its requirement that speech must be likely to incite or produce immediate lawlessness. While hate speech may be extremely provocative when communicated in-person, the chances are extremely remote that such discourse posted online will ever produce acts of unlawfulness with the degree of immediacy necessary to warrant using the "imminent lawless action" test. As Breckheimer (2002) writes:

> [I]t is unlikely that hate speech—or any other speech that proposes immediate violence—on the Internet will ever be punishable under [the "imminent lawless action"] test... . With the "speaker" and "listener" separated as they are on the Internet, it is extremely difficult for the government to show that posting a message will cause the recipient of that message to engage immediately in the proposed violent or illegal act. (p. 1508)

Breckheimer's argument that government is unlikely to find that immediate lawlessness would arise from Internet hate speech is supported by the Supreme Court's failure to find that such a result was likely to be produced by a national television broadcast of the hate speech at issue in *Brandenburg.* Other legal commentators have echoed Breckheimer's sentiment regarding the inapplicability of the "imminent lawless action" test to Internet speech. Thus, regardless of how well-suited it may be for evaluating speech disseminated in the offline world, the "imminent lawless action" test is unsuitable for assessing communications posted online.

## FIRST AMENDMENT TESTS FOR HATEFUL SPEECH THAT ARE APPLICABLE TO HATE SITES

### "Clear and Present Danger" Test

The "group defamation" and "imminent lawless action" tests were intended to apply only to speech thought likely to produce group violence. By contrast, the "clear and present danger" test, as formulated by Holmes in *Schenck*, was designed for application to speech considered likely to create a threat to government. Indeed, *Schenck*, and the other three cases in which the Supreme Court applied Holmes' "clear and present danger" test in 1919 (*Abrams v. United States*, 1919; *Debs v. United States*, 1919; *Frohwerk v. United States*, 1919), involved Espionage Act convictions leveled against persons who disseminated speech that was perceived as threatening the U.S. government's war efforts during World War I. Considering that the standard was intended to apply to speech believed likely to bring about acts that would endanger government, as opposed to groups, the "clear and present danger" test's applicability easily extends to the Internet. The "clear and present danger" test's applicability to the Internet is made even easier by the fact that it provides only a single, general guideline for evaluating speech believed likely to lead to conduct that would threaten government. That the "clear and present danger" standard can, and perhaps should, also be applied specifically to hate sites is discussed in more detail next.

While it is virtually impossible that hateful material posted online could produce large-scale, group violence of any sort, it is very reasonable to assume that the same sort of content is capable of inspiring a small group of individuals to act out violently against government. As witnessed by Timothy McVeigh's bombing of the Alfred P. Murrah Federal Building in Oklahoma City in 1995, words can be of such a nature as to constitute a "clear and present danger" that they will bring about violent attacks on government by a handful of individuals. Although McVeigh's terrorist act is believed to have been inspired by the racist novel, *The Turner Diaries* (Breckheimer, 2002; C. E. Smith, 2002; Tsesis, 2000), which included a fictional account of the bombing of a FBI building, there is no reason to believe that he would not have been motivated by reading the same material online. Yeazel (2002) points to the Oklahoma City bombing as justification for governmental regulation of bomb-making instructions on publicly accessible Web sites, online information which he believes could easily be procured by hate groups and other extremists interested in perpetrating acts of hatred-inspired violence.

Even in the absence of material as obviously threatening as a bomb-making manual, a hate site could still pose a "clear and present danger" to government if its rhetoric was intended to inspire persons to commit acts which would harm government officials or disrupt governmental operations. There is already evidence to suggest that right-wing extremists like McVeigh have been inspired to act as "leaderless resistors" and "lone wolves," persons who independently carry out an ideology's call to violence, after visiting hate sites (C. E. Smith, 2002). Benjamin Smith's two-state shooting spree in 1999 appeared to have been motivated by his frequent visits to a Web site advocating leaderless resistance maintained by the neo-Nazi group World Church of the Creator (C. E. Smith; Tsesis, 2000). Smith killed a black man and an Asian postal employee and wounded nine Jews (C. E. Smith; Tsesis). In that same year, brothers Matthew and Tyler Williams, apparently inspired by lone wolf ideology expressed on white supremacist sites, murdered a homosexual couple in California and torched three synagogues (C. E. Smith). In April 2000 Richard Baumhammers, a person with a strong anti-immigrant sentiment who frequently visited white supremacist hate sites, went on a murder spree after visiting the White Aryan Resistance Web site (C. E. Smith). Baumhammers killed his Jewish neighbor, one black man, one Indian-American man, and two people of Asian descent (C. E. Smith). The actions of these four individuals make it clear that, in certain circumstances, hate sites fuel hate-based violence (C. E. Smith). As Romero (2001) notes, "it would arguably not be much...of a fiction to recognize a causal link between website hate speech and crimes committed by race mongers inspired by such propaganda" (p. 9).

It is apparent that the acts just discussed, including the postal employee murder, were not directed at government. However, the danger of hate sites inciting "lone wolves" to lash out violently against governmental entities seems apparent when one considers the terrorism perpetrated by extremists in the environmental and antiabortion movements. Take for example the Web site maintained by the radical environmentalist group Earth Liberation Front (ELF), which federal law enforcement authorities used to regard as one of the major domestic terrorist threats in the United States (Ackerman, 2003). ELF's site played a vital role in allowing the organization to communicate with those individuals who made up its "leaderless resistance" (Leader & Probst, 2003). Perhaps inspired by the radical environmentalist ideology espoused on the Web site, lone environmentalists claiming to represent ELF committed repeated eco-terrorist attacks on complexes used by government agencies (*The threat of eco-terrorism*, 2002) that they seemingly blamed for failing to support environmentalism. Although the author is unaware of anyone ever being physically harmed as a result of any of these attacks, there has been millions of dollars in property damage (Ackerman, 2003; *The threat of eco-terrorism*, 2002).

However, physical harm—including death—has resulted from acts committed by supporters of the radical antiabortion movement, a movement that also posts its ideology on the Internet, as evidenced by the *PPCW* case. Although the antiabortion violence which prompted the *PPCW* litigation was not directed against government, it is important to remember that, in addition to listing physicians, the Web site at issue in the case included the names of judges and politicians. Thus, it is well within the realm of probability that the same sort of violence that the lone antiabortionists perpetrated against those abortion providers listed on the site would have someday befallen those government officials that were identified as supporting abortion procedures. Likewise, it is also reasonably probable that hate site content may inspire persons to not only carry out acts of violence against individuals who belong to a group which is a target of their hatred, but also against those governmental entities that support the targeted group's interests. For example, it would not be too much of a stretch of the imagination to believe that Baumhammers' strong anti-immigrant sentiment might have eventually led him to exact violence against those government officials who were supportive of immigration rights. Clearly, there is at least some reasonable basis for believing that hate speech posted online could inspire a small group of individuals to commit acts that would threaten government agents, installations, or interests (e.g., police activities).

Yet, while it remains to be seen whether hate site content will incite violence specifically against government, the likelihood of such a result has surely increased as hate groups with antigovernment attitudes have taken to posting their rhetoric online. Indeed, this is an area in which hate sites may pose the greatest danger to government and, as such, the "clear and present danger" test's appropriateness as a guideline for evaluating online speech can be fully realized. Certainly, if the *Schenck* defendants' distribution of several thousand leaflets critical of government could form the basis for applying the "clear and present danger" test, then there must be a rational basis for the test's application to the Internet, where antigovernment rhetoric can be read by millions. To the extent that antigovernment content posted on hate sites may include a call for violence against government, application of the "clear and present danger" test seems particularly appropriate, especially given the Oklahoma City bombing and 9/11.

Of course, it may be argued that even if the "clear and present danger" test is applicable to the Internet, the scope of its application would nevertheless be limited to speech posted online during wartime. Someone making this argument might point out that the standard was developed as the U.S. government was ending its war efforts in World War I, and in formulating the test, Justice Holmes declared that:

> When a nation is at war many things that might be said in times of peace are such a hindrance to its effort that their utterance will not be endured so long as men fight and that no Court could regard them as protected by any constitutional right. (*Schenck v. United States*, 1919, p. 52)

Hence, Holmes' commentary might suggest that he had intended for his "clear and present danger" test only to be applied during times of war. Yet, regardless of when Holmes intended for his "clear and present danger" test to be applied, there have been justices who have indicated their belief that the standard's application, if at all constitutionally valid, was only appropriate during wartime. For example, in his concurring opinion in *Brandenburg*, Justice Douglas wrote, "Though I doubt if the 'clear and present danger' test is congenial to the First Amendment in time of a declared war, I am certain it is not reconcilable with the First Amendment in days of peace" (*Brandenburg v. Ohio*, 1969, p. 452). However, regardless of how wise Douglas' views on the standard may be, the scope of the "clear and present danger" test's application has, in fact, been extended to speech acts that took place in days of peace (e.g., *Wood v. Georgia*, 1962). So, even if the current "war on terrorism" is not a declared war, inasmuch as Congress has never declared it so, it is not determinative of whether the "clear and present danger" test can be used to evaluate hate sites, as the standard has also been applied in peacetime.

Although courts continue to apply the "clear and present danger" standard in hate speech cases (Tsesis, 2000), the test is not typically mentioned in legal commentaries about the regulation of online hate speech. Yet, it is apparent from the analysis provided in this chapter that the Holmesian standard for regulating speech can be readily applied to hate sites. In fact, after concluding that the "imminent lawless action" test is not applicable to the Internet, some legal scholars have suggested that online hate speech regulation is only possible if reliance is placed on a "no-imminence standard" (Komasara, 2002, p. 854), which is essentially the standard provided by the "clear and present danger" test. Thus, contrary to what some scholars may argue, the "clear and present danger" test may well be one of the most appropriate First Amendment standards for assessing the constitutionality of speech posted online, and on hate sites in particular. Although it is the earliest of First Amendment standards, the "clear and present danger" test's general guideline (i.e., no face-to-face speech or imminent lawlessness requirements) and limited focus on speech considered threatening to government make it well-suited for evaluating hate sites. The same characteristics that make the "clear and present danger" test applicable to hate sites, also renders its first progeny, the "bad tendency" test, applicable to hate speech posted online.

**"Bad Tendency" Test**

If the "clear and present danger" standard can be applied to hate speech posted online, then it hardly demands much explanation as to why the "bad tendency" test, an even more relaxed approach to regulating speech (Strossen, 2001), is also applicable to hate sites. Like the "clear and present danger" test, the "bad tendency" test provides a single, general guideline (i.e., no face-to-face speech or imminent lawlessness requirements) for evaluating speech that poses a threat to government, which again some hate site content may reasonably be considered as posing. Furthermore, if there is a reasonable basis for believing that hate speech posted online could potentially pose a "clear and present danger" to government, then it must follow that some hate sites may reasonably be perceived as having a natural tendency to bring about a substantive evil that threatens government. What is more, the "bad tendency" test was formulated in days of peace. So, there is nothing to preclude its application to expressive activity, engaged during days of war—times in which government has historically been even more speech restrictive. Taking the liberty of reapplying and reversing Justice Douglas' "clear and present danger" test logic from his concurrence in *Brandenburg*, if the "bad tendency" test is congenial to the First Amendment in days of peace, then it is certainly reconcilable with the First Amendment in time of a declared war.

Given all these factors, the "bad tendency" test, like the "clear and present danger" standard, is applicable to hate sites. Although some believe the highly speech-restrictive nature of the "bad tendency" test puts it at odds with the First Amendment, no constitutional problems exist if the standard is used only to assess the constitutionality of speech considered to pose a "real and substantial" danger to "organized government." This is precisely the sort of speech that the test was manifestly intended to apply, as shown by Justice Sanford's comments in *Gitlow* and the context in which he made the statements. Furthermore, in the wake of 9/11, courts may be even more inclined to apply the "bad tendency" test to speech considered to pose a real and substantial danger to organized government in the United States. The 20th century would run its course before the Supreme Court would develop another First Amendment test for assessing hateful speech communicated in non-Internet settings that was also applicable to hate speech posted on the Internet.

**"True Threats" Test**

While *PPCW* essentially established that a "true threats" test is relevant in assessing the constitutionality of hate sites, it was decided the year prior to *Black*, the case in which Justice O'Connor devised the Supreme Court's "true threats" standard. However, in the future the "true threats" test used in *PPCW* (and for that matter the one used in *Baker*) will likely be replaced by the standard the Supreme Court formulated in *Black*. Although *Black* involved hate speech in non-Internet settings, the "true threats" test developed by O'Connor in the case is nevertheless applicable to hate sites.

While both of the Supreme Court's "true threats" cases have involved non-mass-mediated communications of hatred, it has long been recognized that individuals can also deliver threats through the media of communication. Indeed, many, if not most, "true threats" are communicated through mediated means, such as a telephone call or mailed letter. This truism is further supported by the simple fact that federal courts have sustained persons' convictions for making "true threats" over the telephone, on television, through the mail, and most recently, via the Internet. Keenly aware that threats can also be communicated through mediated means, O'Connor appears to have formulated her "true threats" test in *Black* so that it could be used to assess the constitutionality of allegedly threatening speech delivered in non-mass-mediated settings and through communication media such as the Internet. The applicability of O'Connor's "true threats" test specifically extends to hate sites given that the two basic guidelines that make up the standard can readily be applied to these sorts of Web sites. First, individuals can just as easily use hate sites for the purposes of communicating a serious expression of intent to commit an act of unlawful violence to a particular person or group of persons to which they bear ill will. Second, those individuals or groups to which threatening messages are directed are no less likely to fear victimization by unlawful violence simply because the communications are posted on hate sites.

Thus, while the "true threats" standard O'Connor formulated in *Black* was used to assess the constitutionality of hate speech communicated in non-Internet settings, the test is clearly written broadly enough that its application can also be extended to hate sites. As such, O'Connor's "true threats" test, the standard relied upon by the nation's highest court, should be adopted by any court asked to determine the First Amendment validity of hate sites.

## Summary

As the U.S. District Court for the Eastern District of Pennsylvania stated in *ACLU v. Reno* (1996), "the Internet...presents unique issues relating to the application of First Amendment jurisprudence..." (p. 830). Therefore, in discussing the possibility of regulating online hate speech it is important to determine how well existing First Amendment law applies to the Internet (Siegel, 1999). This chapter analyzed the six First Amendment tests that the Supreme Court has created for assessing the constitutionality of hateful speech communicated in non-Internet settings to determine their applicability to hate speech posted online. Analysis revealed that only half of these tests are applicable to hate speech posted on the Internet. This is not all that surprising perhaps, since, as Gey (1999) points out, "the ethereal nature of the Internet—which is both everywhere and nowhere—renders obsolete many of the Court's old rationales for regulating speech" (p. 184). The scope and specificity of each First Amendment test was determinative of whether the standard was applicable to hate sites.

Of all the First Amendment tests that the Supreme Court has formulated for evaluating non-Internet communications of hatred, its "fighting words" test is by far the least applicable to hate sites. In *Chaplinsky*, the Court established that the standard was only to be used in evaluating utterances made during a face-to-face encounter, a very specific threshold requirement that cannot be satisfied by the posting of content on the Internet. Similarly, the "imminent lawless action" test is inapplicable to hate sites by virtue of its specific requirement that speech be likely to incite or produce immediate lawlessness, an effect of which online materials are not likely to bring about. The "imminent lawless action" test's applicability to hate sites is further diminished by its scope, as it was intended to apply only to speech thought likely to produce group violence. Group violence is so unlikely to result from online content as to render the "imminent lawless action" standard unsuitable for assessing Internet communications altogether. Given that it has the same scope as the "imminent lawless action" standard, the "group defamation" test is also not applicable to hate sites. Indeed, the validity of the "group defamation" test is altogether doubtful.

Conversely, the scope and specificity of the other First Amendment tests are such that they can be used for evaluating hate site content. The "clear and present danger" and "bad tendency" tests were formulated for use in assessing the constitutionality of speech perceived as a threat to government. Hate speech posted online, like that disseminated offline, could conceivably pose such a threat. The Supreme Court's "true threats" test is equally applicable to hate sites given that it can be applied to threatening speech directed at an individual or a sufficiently small group of individuals, something which is occasionally a component of the hateful rhetoric posted on the Internet. While the Court's "true threats" test is more specific than its "clear and present danger" and "bad tendency" tests, the guidelines provided by each of the standards are general enough that they can be applied in the Internet context, even to content merely posted on hate sites.

Hence, it is the oldest ("clear and present danger" and "bad tendency") and the newest ("true threats") of the Supreme Court's First Amendment tests for determining the constitutionality of hateful speech communicated in non-Internet settings that are also applicable to hate sites. Together with the *Miller* test for obscenity that was identified as being adopted by the Court for use in assessing Internet speech, these standards are useful in assessing the constitutionality of hate sites.

# Chapter 7

# Themes That Appear on Internet Hate Sites

The First Amendment laid the constitutional foundation for protecting the rights of individuals to disseminate hate speech. However, it would be several decades after the amendment's enactment before U.S.-based hate groups were capable of communicating their message of hatred to anything more than a very limited audience. Prior to the development of the postal service and telephone system, all groups wanting to communicate their views to the general public were confined to in-person communication, such as pamphleteering and face-to-face talk. Even when it became possible to communicate by way of mail and telephone, neither mode of communication provided hate groups with an effective means to spread their message to a large audience. After all, mail service and telephone calls only allow for the time-consuming, and perhaps costly, transmittal of one-to-one communications.

Also, hate groups' ability to communicate to larger audiences was not significantly improved by the arrival of the media of mass communication, media which allow for one-to-many transmission of messages (Dimmick, Sikand, & Patterson, 1994). High operating expenses and government-mandated public interest requirements still effectively barred hatemongers from utilizing the media. Currently, the rhetoric of hate groups is usually only conveyed by way of the traditional mass media (i.e., television, radio, newspapers) as part of news coverage about a group's public demonstration. Although in recent years hate groups have been able to utilize public access television as a means of communicating their message of hatred (Harmon, 1991), these stations tend to attract a small audience.

However, just when it seemed that hate groups were spiraling into ever-greater silence, the Internet rose to prominence in the 1990s and reversed that trend. With the Internet, virtually all persons, including hatemongers, have a relatively simple and cost-effective means to communicate to an audience of millions. Realizing the Internet's communication potential, hate groups have increasingly decided to post their rhetoric of hatred online. This chapter identifies the various themes that emerged from the content posted on the publicly accessible areas of the 10 U.S.-based hate sites selected for examination in this study.

## KU KLUX KLAN

### American White Knights of the Ku Klux Klan (AWKKKK)

AWKKKK posted an abundance of content on its Web site. Much of the content on the group's site consisted of articles authored by reputable media figures and organizations. Not adverse to self-promotion, AWKKKK posted and provided links to articles that mentioned the group, its members, and its offline activities. Still, AWKKKK included a lot of original content on its site. In analyzing the content that the group posted to the various hate pages composing its Web site, several themes emerged.

Religion was the main theme that emerged from AWKKKK's online discourse. AWKKKK's site contained a myriad of scripture and religious rhetoric, as well as the text of religiously oriented speeches given by persons like George Washington. Referring to itself as a Christian organization, the group went so far as to claim that the Ten Commandments serve as its "Aryan Code of Conduct." Operating from a Christian perspective, AWKKKK espoused anti-evolution rhetoric and explained why it believes the Christian Bible commands parents to discipline (e.g., spank) their children. The group also expressed its belief that Christian-based religious values are declining in the United States, something which was a frequent topic on the group's site. In addition, AWKKKK commented on how it believes Christians and Christianity are increasingly coming under attack in the United States. The group even had a section on its site entitled "Kristian Kids Korner." The "Kristian Kids Korner" contained scripture, pro-Christian rhetoric, and Bible trivia games, as well as links to externally maintained antiporn, parental control, and religiously oriented Web sites. However, despite claiming to be a Christian group, AWKKKK used profanities on its site.

Racial separatism was another theme that was found on AWKKKK's Web site. The group's reading of the Christian Bible leads it to express support for racial separatism, especially between blacks and whites. On one of its hate pages, AWKKKK claimed that the Christian Bible states that the races should remain separate, and that acts such as interracial marriage are in direct violation of God's will. However, AWKKKK supports racial separatism, especially white separatism, on secular grounds as well. In various places, AWKKKK contended that white separatism is vital to whites' survival because it helps preserve the genetic integrity of the white gene pool. The group claimed, for example, that children born of two white parents tend to be more resistant to disease than are those who have a white and a nonwhite parent. In accordance with its segregationist beliefs, AWKKKK only accepts members of the white Aryan race, which was defined on the group's site as persons with Celtic, Slavic, Nordic, Basque, Germanic, Lombard, or Anglo-Saxon backgrounds. While it supports racial separatism, AWKKKK claims to believe that all races should be accorded equal rights. On its site, AWKKKK stated that no race of people, including whites, is justified in violating another race's rights.

However, while AWKKKK purports to believe that all races should be granted equal rights, the group does not believe that all races are equal. Indeed, AWKKKK repeatedly claimed that whites are the supreme human race on its hate site. The group argued that whites are a superior race because they are mentally superior and because their value system is rooted in Christianity. AWKKKK also provided links to sites maintained by other hate groups (e.g., KKK, neo-Nazi) that share its white supremacist beliefs.

While religion and racially based themes dominated the hate pages maintained by AWKKKK, many of the issues that the group discussed were of a political nature. Since AWKKKK believes that whites are steadily losing their rights, much of its political rhetoric took the form of white advocacy. AWKKKK advocated that whites work to protect their rights by becoming politically active, be it through voting, supporting prowhite causes, or by getting involved in prowhite groups. For example, AWKKKK encouraged whites to vote against any legislator or legislation that would remove the Confederate battle emblem from state flags or properties. The group argued that the Confederate battle emblem is a symbol of white Americans' Southern heritage.

In addition to its white advocacy, AWKKKK devoted considerable space on its site to criticizing the U.S. government and its policies, both foreign and domestic. The group also expressed its politically conservative position on a variety of issues, such as its support for capital punishment and its opposition to abortion, immigration, gay marriage, gun control, and affirmative action. Some of AWKKKK's conservative political rhetoric, however, simply involved criticizing liberals, endorsing conservative politicians, encouraging individuals to vote Republican, or urging persons not to vote for politicians that it believed were not operating in the best interests of conservatives. AWKKKK's political conservatism was also reflected in the fact that it provided links to sites that included conservative commentary, sold conservative publications, or were maintained by politically conservative groups (e.g., National Rifle Association) and news organizations (e.g., Fox News). Other political rhetoric on AWKKKK's site ranged from legal information to statements expressing concern that, as a result of 9/11, the federal government will find it necessary to further restrict Americans' First Amendment freedoms.

Although not nearly as prevalent as its other online rhetoric, AWKKKK also provided some social commentary on its site. Among other issues, the group expressed concerns about the rising prison population in the United States and joked about Americans' obsession with communication technologies, such as pagers and cellular phones.

**Mississippi White Knights of the Ku Klux Klan (MWKKKK)**

In contrast to the site operated by AWKKKK, the hate site maintained by MWKKKK did not contain much content. Moreover, much of the textual rhetoric that MWKKKK did post online was either repetitive, often appearing verbatim on its other hate pages, or just historical information on the Klan and its early leaders, like Confederate General Nathan Bedford Forrest. Yet, several themes still emerged from the rhetoric MWKKKK posted on its site.

Like AWKKKK's site, religion was the main theme underlying the rhetorical discourse on MWKKKK's hate site. On its main page, MWKKKK stated that it stands for, among other ideas, preserving Christianity and the right to practice the Christian faith. MWKKKK expressed its devotion to Christianity on several of the other hate pages that made up its site. As an example, on one of its hate pages the group explained that it no longer "burns" crosses to send a warning, but rather "lights" them as a symbol of its Christian faith. After examining the content of MWKKKK's site, it became clear that among the prerequisites for membership in the group are a belief in Jesus Christ and a willingness to uphold God's commandments.

However, MWKKKK went beyond merely expressing its commitment to the Christian faith to providing its own interpretations of religious writings. MWKKKK maintained a "Bible" page on which it provided its interpretations of what particular verses in the

Christian Bible state about Jesus, impurity, wisdom, self-defense, homosexuality, and race-mixing. The group also maintained a page that included its interpretations of what verses from the Talmud (i.e., one of the Jewish holy books) state about biblical figures, Christians, Christian teachings, how Jews may conduct themselves and should view non-Jews, and so forth. MWKKKK also provided links to other religiously oriented Web sites.

Based on its interpretation of religious texts, MWKKKK believes in racial separatism and the supremacy of the white race, the other major themes to emerge from the rhetoric that the group posted on its site. In particular, MWKKKK believes that white Aryans are God's chosen people, descendant from Adam, and thus should avoid producing offspring with members of other races. One method that the group proposed on its site for segregating the races, and apparently reducing the likelihood of interracial sex, was a three-tiered school system in which parents would have the choice of sending their children to white-only, black-only, or integrated schools. However, when it comes to its own membership, MWKKKK is firmly committed to a segregationist policy of only permitting members of the white Aryan race. Mirroring the definition that AWKKKK posted online, the white Aryan race was defined on MWKKKK's site as persons whose background is Anglo-Saxon, Germanic, Nordic, Basque, Lombard, Celtic, or Slavic.

In addition to the themes of religion, racial separatism, and white supremacy, there were several issues of a political nature addressed on MWKKKK's hate site. As an example, MWKKKK indicated that it opposes laws limiting Americans' freedom of speech. The group also expressed its opposition to laws legalizing gay marriage by providing a link to www.nogaymarriage.com on its main page.

However, MWKKKK's political rhetoric was witnessed most notably in the group's call for the state of Mississippi to enact additional child molester legislation, an issue the organization devoted a significant portion of its site to discussing. MWKKKK explained its reasons for believing Mississippi's then-current child molestation laws needed to be enhanced, provided its proposal for enhancing those laws, and made repeated requests for persons to assist the group in petitioning the state to adopt its proposal. Indeed, MWKKKK's site contained a petition calling for enhanced child molestation laws, which the group urged persons to print, distribute for signatures, and return to the organization so that it could forward the request, along with other petitions, to the Mississippi legislature. MWKKKK also provided links to Web sites operated by organizations supporting sex offender registration and community notification, the two aspects of Mississippi's then-current child molestation legislation that the group believed were in need of enhancement.

Several minor themes were also addressed on MWKKKK's hate site. The group stated that it is interested in the preservation of the Mississippi state flag, as well as Western culture and civilization. Viewed in context, however, even MWKKKK's preservationist interest in these areas seems to be just an extension of its white supremacist beliefs. Mississippi's state flag includes a Confederate battle emblem which is often regarded as a symbol of white supremacy, and the term "Western," when used to describe a culture or civilization, denotes what is typical in the predominantly white regions of Europe and the United States.

## Neo-Nazi

### National Socialist Movement (NSM)

Unlike MWKKKK's site, the hate site maintained by NSM contained a plethora of content, more than any other Web site examined in this study. NSM's site consisted of several hundred hate pages. While most of these pages contained pro-Nazi and prowhite rhetoric, many simply included information about NSM's merchandise, members, membership policies, and offline activities, namely its public demonstrations. In addition to posting a surplus of content on its site, NSM provided links to a profuse number of externally maintained Web sites, including those operated by conservative groups, pro-children associations, nonprofit organizations (e.g., Red Cross), and professional news media, both in and out of the United States. Like

AWKKKK, NSM promoted itself by posting, and providing links to, articles that discussed the group, its followers, and its offline activities. Composed of such a vast array of content, NSM's site inevitably contained several different themes.

As one might expect from a Web site operated by neo-Nazis, white supremacy was the main theme of NSM's hate site. Not only did NSM repeatedly express its loyalty to white supremacists like Adolf Hitler and other Nazi figures from the past, the group indicated that it shares their belief that the white race is superior to all other races, particularly in terms of intelligence. For example, on one of its hate pages NSM claimed that whites are responsible for creating all things of true importance to society. Believing in the supremacy of the white race, the group indicated that it feels that whites should also avoid producing biracial offspring with nonwhites. NSM posted a variety of other white supremacist rhetoric on its site. Much of this rhetoric suggested that white Americans need to start standing up and fighting for the white race and its rights, rights which the group claimed are slowly becoming eradicated in the United States. NSM also promoted and reported on the offline activities of other white supremacist groups, both neo-Nazi and KKK. In addition, NSM provided links to articles about white supremacy and to Web sites operated by other white supremacist groups, as well as various companies that distributed white supremacist merchandise, such as books, video games, and music recordings.

Along with white supremacy, a major political theme ran through NSM's site. For example, the group frequently expressed its support for upholding the Constitution and its respect for early American political figures, such as George Washington. The heavy political orientation of NSM's site was probably due to the fact that the group considers itself a political party, albeit one solely dedicated to serving the interests of whites. Indeed, NSM announced that it intended to field candidates in the 2008 presidential election on its site. Although the general public is unlikely to ever recognize it as a legitimate political party, NSM nevertheless provided its political platform. As indicated on the group's site, NSM's political platform includes its opposition to abortion, gun control, and nonwhite immigration, as well as its plans for reforming the nation's tax, welfare, education, and health care systems.

Most of the political rhetoric on NSM's site involved criticism of the U.S. government. Among other complaints, the group criticized the U.S. government for failing to stop illegal immigration, spending taxpayer money on wars in foreign countries instead of on national defense, and its relationship with the United Nations. NSM believes that the United Nations does not operate in the best interests of Americans. NSM's site also contained critical commentary about U.S. governmental policies that the group views as harmful to white Americans, such as affirmative action job programs for nonwhites.

However, the most frequent criticisms that NSM leveled against the U.S. government on its site were accusations that its domestic and foreign policies are geared more toward the interests of Israel and Jewish Americans than to the United States and non-Jewish white Americans. As an example, the group explained that it opposes the war in Iraq because it believes that the conflict was solely intended to help make Israel more secure, not the United States. In fact, NSM claimed that most of the U.S.-led wars in the Middle East are fought on behalf of Israel. NSM also indicated its opposition to U.S. support for Israel, which the group accused of committing several atrocities in the Middle East, such as killing masses of Palestinians.

Along with its own political rhetoric, NSM posted statements made by American and non-American political figures, both living and deceased, as well as various reports (e.g., al Qaeda activities, hate crimes statistics) issued by governmental agencies. The group also provided links to various government (e.g., Homeland Security, Census Bureau) and politically oriented Web sites, such as those lobbying for anti-immigration legislation or an end to the war in Iraq.

NSM's site also had a minor religious theme. In addition to posting a variety of pro-Christian rhetoric on its site, NSM maintained a Web page that contained several verses from the Talmud that it claimed demonstrate that Judaism is essentially an evil religion that teaches Jews to hate non-Jews. The group also posted content relating to religion generally, such as the issue of religious freedom. Among other criticisms, NSM complained about what it perceives as religion being forced out of American society. The group also cited passages from the Bible, although sometimes in an effort to show that Jews are evil and set on world domination. In spite of its pro-Christian rhetoric, however, NSM used profanities at numerous occasions on its site.

## Racist Skinhead

### Tualatin Valley Skins (TVS)

Like NSM, TVS posted a wealth of content on its hate site. TVS's Web site consisted of e-mails sent to the group, as well as the text of several books available in public libraries, such as Sun Tzu's *On the Art of War*, Hitler's *Mein Kampf*, and the racist novel *The Turner Diaries*. Seemingly obsessed with self-promotion, TVS used its site to advertise its various offline activities. Among other items, the group furnished schedules for its public access television appearances, and posted or provided links to news stories about acts committed by its members, like distributing racist literature. TVS also posted a variety of articles created by professional media organizations in an effort to support its own positions. In some cases, the group engaged in a sort of "journalistic revisionism" by inserting its own commentary at those points in an article where it believed there was a need for correction or clarification.

However, while most would have questioned the accuracy of the majority of TVS's online commentary, there was no question that the group embraced a white supremacist ideology. On its site, TVS repeatedly expressed its support for Hitler, the 20th century's most notorious white supremacist, and his belief that the white race is superior to all other races. For instance, the group expressed the view that "White is right" and all nonwhites are genetically inferior "subhuman debris." Inspired by notions of white supremacy, TVS indicated that it was dedicated to fighting for—and recruiting others to fight for—the preservation of the white race through legal means, like distributing prowhite literature. Indeed, the group endorsed the practice of disseminating flyers that contained comments suggesting that whites need to avoid mixing with other races to save their own race from extinction. One of TVS's hate pages even provided suggestions for persons interested in spreading a message of white supremacy, such as designing a hate site, forming a white supremacist band, appearing on public access television stations, and distributing prowhite literature.

TVS also believed (and even seemed to fantasize) that someday a racial war would take place in the United States which would require white Americans to physically fight for their continued existence. As such, the group recommended that its fellow white supremacists get physically fit and become adept in hand-to-hand combat and the use of guns. TVS also posted information regarding members in, and the offline activities (e.g., rallies) of, other white supremacist groups, like NSM and the Imperial Klans of America, and provided links to these organizations' Web sites.

Although white supremacist rhetoric made up the bulk of TVS's site, other themes did emerge from the content that the group posted online. For one, TVS provided some social commentary, albeit commentary that was largely rooted in the organization's white supremacist ideology. As an example, the group expressed concern that the large Jewish presence in the American media and entertainment industries was causing Americans to be provided with a biased and incomplete picture of issues and events in the United States and abroad. To justify its fears, TVS listed the names of Jewish-owned media in the United States, as well as those of several Jews that were currently managing or were employed in America's media and entertainment industries.

Additionally, TVS's site contained a great deal of political rhetoric, although nowhere near as much as the Web site maintained by its NSM counterparts. For instance, the group posted several comments made by political figures, past and present. However, most of the political rhetoric that TVS posted online was its own. As an example, TVS offered legal advice to persons wanting to support the group's political objectives. In addition to self-serving political rhetoric,

TVS offered a lot of critical commentary about the U.S. government and its actions, such as its decision to go to war in Iraq without the support of the United Nations. The group's political rhetoric also included criticisms of foreign governments, especially Israel's government. TVS claimed that Israel is a terrorist state responsible for terrorizing Palestinians, attacking the USS Liberty, selling weapons to enemies of the United States, and causing 9/11. As indicated by the group's Web site address, www.nukeisrael.com, TVS was in favor of the destruction of Israel.

## CHRISTIAN IDENTITY

### Orange County Assembly of Christ (OCAC)

Of all the hate sites examined in this study, the Web site operated by OCAC contained the least rhetorical discourse. Roughly two-thirds of OCAC's Web site was devoted exclusively to advertising (i.e., titles, short descriptions, photos) the various books that the church had for sale, which were of a religious, historical, political, scholarly, and scientific nature. Despite the lack of content that OCAC posted on its site, a couple of themes consistently emerged from the church's online rhetoric.

Not surprisingly, religion was the main topic discussed on OCAC's site. Upon visiting the site's main page, persons were immediately presented with verses from the Bible, the church's religious views and mission, and various links that connected to Christian-oriented Web sites, such as www.ChristianBibleStudy.org. Religious content of this sort appeared throughout the group's site. OCAC also used its Web site as a vehicle to promote the church's offline religiously oriented events and activities, such as worship services, missionary programs, public protests, and the distribution of religious literature in its community.

OCAC's site also included advertisements for other churches' offline religious events and activities, as well as information about Christian Identity programs and sermons broadcast by other parties via shortwave radio and over the Internet. OCAC provided links to external sites where audio from some of these programs and sermons were posted. The church also provided links to some externally maintained Web sites that it marketed as being for "Christian Kin," but which appeared to be completely secular sites (e.g., debt consolidation) that were simply being advertised to generate revenue. Generating revenue seemed to be among the main purposes OCAC had in posting its site.

However, while OCAC appeared to be trying to establish a union with the various proponents of the Christian Identity theology, the church did not believe such unity should exist between whites and the various other races of people inhabiting society. Like some of the white supremacist groups examined in this study, OCAC's interpretation of the Christian Bible led it to believe that whites are a supreme race of people. On one of its hate pages, OCAC claimed that whites are "true Israelites" descended from Adam, commanded by God to remain separate from other races. In accordance with its religious beliefs, the group billed itself as both a "Christian Anglo-Israel Congregation" and "A Church for White People" on its site. The fact that OCAC indicated that its weekly church services were "by invitation only" seemed to be an obvious effort by the group to make sure that Jews and nonwhites did not attempt to become part of its congregation.

Not only did OCAC's beliefs in racial separatism and the supremacy of the white race mirror those held by many Klan groups, the church indicated that it aligned itself with some KKK organizations. For example, the group expressed its appreciation for the White Camelia Knights' "religious teachings" and the Imperial Klans of America's establishment of a scholarship fund for white Christian children.

## Black Separatist

### Nation of Islam (NOI)

While OCAC's hate Web site was operated by whites, religiously oriented black hate groups also have an online presence, as evidenced by the site maintained by the black separatist group NOI. However, unlike OCAC's site, NOI's Web site was very businesslike, consisting mainly of press releases issued by the organization dating back to 1996 on topics ranging from Michael Jackson to the Middle East. Most of these releases simply provided factual information (e.g., dates, times, places) about NOI events (e.g., conventions, Million Man March), activities (e.g., tours), press conferences, and speaking engagements (e.g., live Internet chat sessions). In addition to press releases, NOI's site contained a great deal of information about the group's history (e.g., founders, leaders, mosques, programs), as well as several links to *FinalCall.com News*, the organization's online newspaper.

Amidst the businesslike design of NOI's site, religion emerged as the main theme of the group's online rhetoric. NOI's religious rhetoric permeated virtually every area of its site, be it in the form of prayer, blessings, cited scripture, biblical interpretations, or warnings to persons not to commit acts that would cause them to incur the wrath of God, whose name is "Allah" in the religion of Islam. In fact, written transcripts of some of the religious speeches given by NOI members were posted on the site. Contrary to the religious belief held by MWKKKK, NOI believes that blacks, not whites, are God's chosen people. On its site, NOI argued that blacks most accurately fit the description of a group that God would have chosen. NOI also provided links to other religiously oriented sites, including ones focusing on the Islamic religion.

Intertwined with the religious rhetoric on NOI's site was a large amount of political speech. NOI weighed in on a number of political issues, such as expressing its opposition to the war in Iraq, corporate influence in government, and the U.S. military conducting bombing exercises in Puerto Rico against the wishes of many of its citizens. Being an Islamic group, NOI was particularly critical of actions taken by the U.S. government that affect the Arab and Muslim regions of the world. For example, NOI expressed its opinion that the U.S. government has needlessly conducted military operations in various Arab and Muslim countries, like Iraq. The organization also expressed concerns that the U.S. government will someday restrict the civil rights of Arabs and Muslims living in America.

However, most of the political speech on NOI's site consisted of the group's commentary on the U.S. government's treatment of blacks throughout the world, and in the United States in particular. The organization was extremely critical of the racially discriminatory policies that the U.S. government has had in the past, namely those permitting black slavery. NOI also commented on racially discriminatory policies that it believes are still being practiced in America as a means of oppressing specific members of the black community, and blacks in general. For example, the group claimed that the U.S. government continues to deny its black citizens freedom and equal justice by allowing them to be subject to police brutality and a white legal system that places numerous blacks in prison and on death row. The group also argued that black Americans should be exempt from taxation until it is satisfied that they have been accorded rights equal to those of white Americans.

In addition to claiming that there are several antiblack governmental policies still at work in America, NOI alleged that the U.S. government has secretly engaged in a variety of illegal antiblack activities over time. The group claimed that, among other acts, the U.S. government has sold drugs in black communities in order to finance wars and played a role in the assassinations of Malcolm X, Martin Luther King, Jr., and other prominent black leaders from the 1960s. NOI also accused the government of having engaged in a campaign to destroy its organization during the 1960s. To support this allegation, the group posted several FBI counterintelligence reports from the 1960s in which the government expressed an interest in neutralizing the activities of black nationalists, including those of NOI and its leaders. NOI's site contained links to other sites that examined the FBI's efforts to neutralize political dissidents.

Because of white Americans' past enslavement of blacks, and the antiblack policies and activities that the group believes the U.S. government has endorsed and practiced, NOI called for racial separatism, another theme that commonly appeared in the organization's online discourse. In particular, the group called for the U.S. government to right past and current wrongs by establishing and maintaining (e.g., equipment, supplies, buildings) a separate, but equal, territory for black Americans that are descendant from slaves. Although NOI preaches peace, its online rhetoric indicated that the organization does not believe peace can be achieved between blacks and whites, and therefore it feels racial separatism is the only way for the two groups to peacefully coexist.

**New Black Panther Party (NBPP)**

Another black separatist group with an online presence that has close ties to NOI is the ultramilitant NBPP. In fact, NBPP follows a "10 Point Program" (e.g., reparations, separatism) adapted from that followed by NOI. Like NOI's site, NBPP's Web site included a lot of purely factual information about the group's history (e.g., programs, leaders) and its offline activities (e.g., dates, times, locations), such as its public protests. Unlike NOI's Web site, however, NBPP's Web site contained an abundance of content, much of which consisted of articles from reputable media organizations (e.g., Associated Press), external Internet sites (e.g., Yahoo!), and other problack groups (e.g., Black America Today). Many of these articles discussed black persons or issues of relevance to blacks, like the economic and educational statuses of black Americans.

Another way in which NBPP's Web site differed from NOI's site was in terms of its primary theme. Unlike NOI, NBPP is not a religious organization. As such, NBPP's site only contained a scant amount of religiously oriented content, like an article about the "Black Madonna" and the group's claim that blacks of African descent are the true Semites. Instead, the majority of the content on NBPP's site was politically oriented, just as the group considers itself to be politically oriented.

The political content on NBPP's site ranged from the transcripts of speeches given by political figures (e.g., Lincoln's Emancipation Proclamation speech) to the group's own political commentary. Like many of the hate groups in this study, NBPP commented on the activities of foreign governments. As an example, the group criticized Israel for some of its actions done in opposition to the Palestinians. NBPP also commented on the U.S. government's foreign policies. Among other sentiments, the group expressed opposition to the U.S.-led war in Iraq and explained why it believes that the U.S. government's foreign policy, especially its support for Israel, provoked 9/11. However, most of the political rhetoric on NBPP's site dealt with the U.S. government's domestic policies, particularly as they relate to blacks. Like NOI, NBPP was extremely critical of the racially discriminatory, antiblack policies that the U.S. government sanctioned in the past—in particular the legalization of black slavery—and those which the group believes are still being practiced today.

When it came to discussing past wrongs that the U.S. government perpetrated against blacks, NBPP's political rhetoric primarily dealt with the issues of slavery and slavery reparations. As indicated on its site, the group believes that the U.S. government should issue a formal apology for, or an acknowledgment of, the fact that slavery was once permitted in America. Like NOI, NBPP indicated that it also believes that all African-Americans are entitled to reparations (e.g., cash, land) from the U.S. government because their ancestors were subjected to the slave trade, slavery in America, and years of postslavery discrimination in the United States. Along with its own proreparations rhetoric, NBPP provided links to sites maintained by other groups that believe African-Americans are entitled to slavery reparations. Like NOI, NBPP also posted several FBI documents from the 1960s in which the government indicated an interest in eradicating black nationalists, like the original Black Panther Party, as further proof of the antiblack policies that the U.S. government has endorsed in the past.

As for the current U.S. government, NBPP commented on a variety of its domestic policies. It became clear after examining NBPP's online rhetoric that the group believes most of the U.S. government's domestic policies are antiblack, especially in the areas of criminal justice and law enforcement. With respect to criminal justice, NBPP explained that it believes that the U.S. government has, among other actions, discriminatorily applied the death penalty to blacks and wrongfully imprisoned many blacks for purely political reasons or by denying them a right to a fair trial (e.g., all-white juries). With regard to law enforcement, NBPP discussed how it believes that the U.S. government has allowed blacks to be subjected to murder, brutality, harassment, and racial profiling by police. In fact, the group devoted a considerable amount of space on its site to discussing police mistreatment of blacks, including several high-profile incidents, such as Rodney King's beating at the hands of LAPD officers.

Because of its resentment of America's legacy of black slavery, and what it perceives as ongoing government-sanctioned discrimination against blacks in the United States, NBPP calls for racial separatism, another theme that consistently emerged from the content of the group's site. NBPP's online rhetoric on racial separatism closely paralleled that found on NOI's site.

The scathing criticisms of the U.S. government that NBPP posted online were but one part, albeit the most significant part, of a larger anti-American theme that ran through the group's site. As indicated by the rhetoric that the group posted on its site, NBPP believes that not only is the U.S. government generally antiblack, but so is the rest of American society. This belief stems from the fact that the group believes whites, who hold a majority status in most sectors of American society, are generally racist and do everything in their power to oppress blacks. NBPP's anti-American sentiment, as well as its belief that white Americans are generally racist, was clearly reflected in its use of the term "Amerikkka" to refer to the United States on its site.

However, while NBPP devoted much of its site to expressing its belief that certain individuals and institutions are antiblack, the group also posted a variety of problack rhetoric online. Along with honoring various blacks, both living and deceased, NBPP discussed various ways in which it believes black Americans can become proactive in achieving or retaining "Black Power" in the United States. Among other ideas, the group expressed its belief that blacks need to work toward reducing black-on-black crime, redeveloping black neighborhoods, and controlling all aspects (e.g., economic, education) of life in their communities. In attempting to promote "Black Power" on its site, NBPP provided links to sites maintained by blacks, black businesses, and other problack groups (e.g., NOI). The group also provided links to sites carrying merchandise about blacks (e.g., books on amazon.com) or featuring black artists, such as a site selling hip-hop music albums.

## NEO-CONFEDERATE

### League of the South (LOS)

LOS's hate site had content dating back to 1994. Like NSM's Web site, LOS's site contained an abundance of content and a profuse number of links to external Web sites, including those maintained by educational institutions, nonprofit organizations (e.g., The Cato Institute), and professional news media, both in and out of the United States. Aside from that, LOS's hate site was quite unlike any of the others examined in this study. Much of the content on LOS's Web site consisted of scholarly essays written by individuals such as university professors and persons with advanced college degrees, like Dr. J. Michael Hill, the group's president.

However, LOS's site was not all about scholarship. Like other hate groups in this study, LOS used its Web site for self-promotion. On its site, LOS posted newspaper articles about its group and provided information on its organization and how people could support its agenda. The group also promoted, and reported on, its offline activities, such as its national conferences, private gatherings, and public demonstrations. LOS also used its site to promote, and report on, the online (e.g., newsgroups) and offline activities of its members.

Regardless of what content LOS posted online, it became clear that the group was committed to promoting the culture and prosperity of the South, as well as the interests of American Southerners that it considered among its constituency. Indeed, LOS's pro-South sentiment was the primary theme of its site.

One of the main ways in which LOS expressed its pro-South attitude was by affectionately discussing the "Old South" and Southerners of yesteryear, such as Thomas Jefferson. However, while LOS devoted a great deal of space on its site to fondly discussing Southern history, the group's pro-South sentiment was also reflected in its online commentary regarding current events taking place in the South and elsewhere in the country. For instance, LOS, which dedicates itself to preserving "Southern symbols," frequently discussed why it opposes the ongoing movement to remove the Confederate flag from public properties in the South and from American society in general. In various places, the group argued that the Confederate flag is more a symbol of the South than of hate. LOS also spoke out on behalf of Southerners themselves on its site. As an example, the group condemned the use of derogatory expressions, like the term "redneck," to refer to Southerners. What is more, LOS called for persons to boycott businesses that it felt were discriminating against Southerners in one way or another (e.g., refusing to sell Southerners' products).

Other ways in which LOS expressed its pro-South sentiment were by promoting Southern businesses and events (e.g., festivals), as well as advertising books and movies about the South or Southerners. The group also expressed its pro-South sentiment by selling Southern-themed merchandise (e.g., music CDs) and media artifacts (e.g., videos) that dealt with the South, Southern history, or well-known Southerners. LOS also promoted the offline activities (e.g., meetings) of other pro-South groups. In addition to its own pro-South rhetoric, LOS further promoted the South by providing links to various pro-South and Southern-oriented sites, like ones where persons could download Southern-style music (e.g., Southern folk music).

Most of LOS's pro-South rhetoric, however, was political in nature. For one, the group endorsed certain politicians and legislation that it regarded as being pro-South. At the same time, LOS was critical of politicians and legislation that it regarded as being anti-South. LOS also frequently criticized the Democratic and Republican parties, as well as the U.S. government, which the group referred to as "the American Empire" on its site. The group accused each of these entities of failing to serve the best interests of Americans generally, and Southerners specifically.

LOS's disenchantment with the two major American political parties led it to call for the formation of a pro-South third party on one of its hate pages. The group even explained what it believes a pro-South party should have as its political platform. As for the U.S. government, LOS claimed that "the American Empire" has worked against the interests of Southerners by, among other deeds, failing to adhere to the Constitution, consistently curtailing Southern states' rights, and promoting the removal of Southern symbols, like the Confederate flag. Convinced that the U.S. government is not sufficiently committed to serving the interests of Southerners, LOS commented on why it believes that the Southern states should consider seceding from the United States, a major political theme that emerged from the group's online discourse. In one section of its site, LOS even argued that the U.S. government owes Civil War reparations to the South.

However, not all of LOS's political rhetoric was related to its support for the South and Southerners. The group offered its position on a variety of other political issues, such as its support for tax reform and local self-government and its opposition to gun control, gay marriage, the Patriot Act, and the war in Iraq. In fact, much of LOS's political rhetoric was geared toward advocating the rights of white Americans. The group argued several times on its site that white Americans are increasingly being subjected to reverse discrimination by government in the United States. For example, LOS denounced race-based preference programs on the grounds that they unfairly discriminate against whites. LOS also provided critical commentary regarding the U.S. government's foreign policies, like its support for Israel, and discussed political issues confronting other countries, such as Canada. In addition, LOS's site included quotes from past political figures, the text of historic political documents, and links to politically oriented Web sites, such as those calling for anti-immigration legislation.

Along with political content, LOS's site contained a variety of religious rhetoric, most of it pro-Christian. The group made repeated references to, and often engaged in lengthy discussions about, God, Christianity, and Christian values. Some of LOS's religious rhetoric, however, involved denouncing activities that the group perceives as being anti-Christian, such as abortion, infidelity, and homosexuality. However, most of the group's religious rhetoric expressed concern that the U.S. government is limiting, or failing to respect, Americans' religious freedoms, particularly those of Christian Americans. LOS's site also contained a variety of verses from the Christian Bible, religiously oriented statements made by historical figures, and links to Christian-oriented Web sites.

In addition to political and religious rhetoric, LOS's site contained some social commentary. For example, the group commented on what it perceives as the U.S. news media's failure to aggressively report on hate crimes that have been perpetrated against whites in American society. LOS also commented on affairs taking place in societies outside of the United States.

## Other Hate Groups

### Council of Conservative Citizens (CCC)

A group which is ideologically in tune with LOS is the ultraconservative CCC. However, unlike LOS's site, CCC's Web site only contained a modest amount of content, and much of that was simply factual information (e.g., dates, times, costs, purpose, addresses) about the group's newspaper, chapters, meetings, members, membership policies, call-in radio show, and activities in public settings.

Like LOS's site, however, CCC's Web site was largely composed of content that was political in nature. The group offered timely political commentary about decisions made by government officials such as President George W. Bush, and even provided contact information for government offices and officials, as well as links to various politically oriented Web sites. Remaining true to its name, CCC also offered conservative views on a variety of political issues. Among the politically conservative themes that emerged from CCC's site were the group's opposition to gun control and restrictions on religious speech, as well as its criticism of the U.S. government's use of taxpayer dollars on programs such as welfare. CCC's site also contained links to Web sites that offered information from a conservative political perspective.

However, most of CCC's online political rhetoric took the form of white advocacy. CCC prides itself as being a racially conscious group that advocates and promotes the rights and causes of white European-Americans, especially those subscribing to a conservative political ideology. On its site, CCC spoke out against government programs and policies that it perceived as fostering a sort of reverse discrimination against white European-Americans, like racial profiling, race-based preference programs (e.g., affirmative action), and lax immigration policies.

With regard to racial profiling, CCC claimed that government agencies apply a double standard that permits whites to be profiled as criminals, but not nonwhites. The group argued that race-based preference programs bestow preferential treatment on nonwhites and other preferred minorities at the expense of white European-American taxpayers. As for immigration, CCC contended that masses of nonwhite, Third World immigrants have been allowed to enter the United States, legally and illegally, for so long that they now pose a real threat to the rights of Americans generally, and white European-Americans specifically. The group cited incidents of antiwhite violence that were supposedly perpetrated in other countries as evidence that the growing nonwhite population of the United States threatens the safety of white Americans. Therefore, CCC was ultra-critical of governmental immigration policies, especially as they related to illegal immigrants (e.g., amnesty plans). The group also expressed its extreme displeasure with local, state, and federal government bodies and officials that had not pushed for anti-immigration legislation. Not only was nonwhite immigration into the United States the most common issue addressed on CCC's site, the group also provided links to Web sites maintained by anti-immigration and prowhite organizations.

Like LOS's Web site, CCC's site contained a variety of religious rhetoric, most of it from a pro-Christian perspective. The group expressed a concern for what it perceives as a decline in moral standards and Christian values in the United States. As evidenced by its online rhetoric, CCC believes the decline in American standards of decency is reflected in what it feels is a growing acceptance of homosexuality and pederasty (i.e., nonfamilial man-boy sexual relations) in the United States. As for declining Christian values in the United States, the group pointed to incidents such as Alabama Judge Roy Moore being removed from the bench after refusing to remove a Ten Commandments monument from the courthouse where he worked. However, not all of the religious rhetoric on CCC's site had a negative tone. For example, CCC explained that it believed the extreme popularity of Mel Gibson's *The Passion of the Christ*, a movie that the group endorsed, indicated that many Americans were yearning to fill their spiritual void. CCC's site also included links to various pro-Christian Web sites.

White supremacy also emerged as a theme of CCC's hate site. CCC implied that it believes whites are a superior race of people by continually suggesting that the existence of nonwhites in the United States has somehow diminished white European-Americans' quality of life. If that were not enough, the group posted content on its site explicitly stating its belief that Western civilization has flourished because it was designed by whites (e.g., Greeks, Romans, Celts, Anglo-Saxons) with "superior genes."

Another minor theme that emerged from CCC's hate site was the preservation of Southern heritage. Like LOS, CCC explained that it was committed to fighting for the preservation of what the group regards as symbols of Southern heritage, such as the Confederate flag and the Mississippi state flag. On one of its hate pages, the group argued that if "the liberal establishment" is left to its own devices that it will do away with many symbols of Southern heritage, a heritage which CCC feels rightfully belongs to white Americans. Hence, the interest that CCC expressed on its site for preserving Southern heritage was directly related to its white advocacy.

**Jewish Defense League (JDL)**

Although few in number, there are also some U.S.-based hate sites operated by Jews, such as that maintained by JDL. A militant Jewish group whose motto is "Never Again," a slogan which represents the group's commitment to helping ensure that there is never another Holocaust, JDL billed itself on its site as "the most controversial, yet the most effective, of all Jewish organizations" (Jewish Defense League [JDL], n.d.). Dedicated almost exclusively to Jewish issues and problems, JDL used its site as a forum to discuss topics that the group believed were important to Jews, to speak out on behalf of Jews, and to condemn persons and acts it regarded as being anti-Jew. For example, JDL reported on the activities of anti-Semitic groups and individuals that it believed posed a threat to Jewish Americans. On a more positive note, JDL's site had several pages devoted to promoting Jewish causes and Jewish-run businesses, as well as to honoring the accomplishments of various Jews, such as rabbis, entertainers, professional athletes, World War II heroes, and Holocaust survivors. The group also promoted businesses and persons it regarded as being pro-Jew. Simply put, the main theme of JDL's site was pro-Jew.

As part of its pro-Jew sentiment, JDL posted some religiously oriented content on its site, most of which related to Judaism. JDL cited scripture from the Torah, discussed the Jewish religion and its history, posted essays written by Jewish rabbis, encouraged others to embrace its pro-Judaism attitude, and so forth. The group also expressed concerns that Judaism was under attack in the United States and criticized the efforts of some Christians to convert Jews to Christianity. In addition to discussing Judaism, JDL cited scripture from the Koran (i.e., the holy book of Islam) and the Christian Bible, although usually only those verses which the group regarded as being anti-Semitic in nature.

JDL's pro-Jew attitude was also evidenced by the political rhetoric that the group posted throughout its site. Reflecting its status as an international Jewish organization, JDL commented on issues of political significance facing Israeli Jews. For example, the group discussed the Israeli-Palestinian conflict and why it believes Israel's Jewish community is entitled to the disputed areas. In addition, JDL frequently remarked on how it believed decisions made by the U.S. and Israeli governments would affect Jews living in Israel. As an example, JDL criticized leaders from both countries for supporting the Israeli-Palestinian "peace process," something which the group believes will have disastrous consequences for Jews in Israel. JDL also devoted a lot of space on its site to criticizing U.S. government policies, agencies, and officials that it regarded as being anti-Jew or anti-Israel. For instance, the group's site contained a statement suggesting that the U.S. government is anti-Jew because it caters to Arab countries, in particular Saudi Arabia. Much of JDL's political rhetoric was geared toward getting persons who are pro-Jew and pro-Israel to become politically active. As an example, the group encouraged persons to lobby politicians to enact pro-Jew and pro-Israel legislation. JDL's site also included the text of Congressional bills, contact information for government agencies and officials, as well as the court transcripts of cases in which JDL members have attempted to prohibit local governmental bodies from leading Christian prayer during their meetings.

While JDL used its site to encourage pro-Jewish political activism, its pro-Jew attitude was most clearly reflected in its attempts to promote pro-Jewish social activism. Already socially active itself, the group encouraged others to become pro-Jew activists in society in anyway they could, be it through boycotts, petitions, public protests, phone calls, letter-writing campaigns, or some other means. JDL also called for people to join it in boycotting and/or protesting against certain businesses, organizations, and individuals (e.g., comedians, store owners) that the group accused of being anti-Jew or contributing to anti-Semitism in some way. In most cases, JDL posted contact (e.g., addresses, telephone numbers) and location (e.g., maps, directions) information for those businesses, organizations, and individuals it had targeted for protest. JDL also requested that persons participate in its pro-Jewish counter-demonstrations against anti-Semitic groups (e.g., neo-Nazis) that were planning to rally in public places. In addition to posting information (e.g., dates, times, maps, directions) about its own pro-Jewish offline activities in society, JDL provided information on public demonstrations held by other pro-Jew and pro-Israel groups.

JDL supplemented its own online rhetoric by posting a variety of newspaper, Internet-only media, and Jewish journal articles that discussed topics of interest to Jews, like Israel and anti-Semitism. What is more, JDL provided links to Jewish-oriented articles on mainstream media and Jewish journal Web sites. The group also provided links to sites that were operated by Jews, that were pro-Jew or pro-Israel in nature, and that dealt with issues of relevance to the Jewish and Israeli communities, such as the Holocaust.

## Summary

Like millions of Americans, hate groups based in the United States have increasingly taken advantage of the Internet's communicative power. Where once there were few, now there are several hundred online hate sites operated by U.S.-based hate groups. This chapter focused on the thematic content appearing on the publicly accessible areas of a sample of 10 such hate sites. Included in this sample were Web sites operated by Ku Klux Klan (2), neo-Nazi (1), racist skinhead (1), neo-Confederate (1), Christian Identity (1), black separatist (2), and "other" hate groups (2). The two hate sites in the "other" hate group category consisted of a Web site maintained by a Jewish organization and one operated by a group of white conservatives. Several themes emerged from the content posted on these 10 hate sites.

Some of the major themes that emerged were found only on particular hate sites. One of the two black separatist hate sites (NBPP) was unique in that it contained both a major anti-American and problack theme. The neo-Confederate Web site was the only site to contain a major pro-South theme and to have the issue of Southern state secession from the United States consistently emerge from the rhetoric posted on it. Likewise, the site operated by the Jewish organization was the only hate Web site to contain a major pro-Jew theme and to have an emphasis on pro-Jewish social activism. However, most of the major themes that emerged were common to at least half of the hate sites.

One of the most common themes to appear in the rhetorical discourse of the sampled Web sites was that of white supremacy, which is not all that surprising since the majority of the hate sites examined in this study were operated by whites. White supremacy was the main theme of the neo-Nazi and racist skinhead Web sites, and a major theme on the Christian Identity site, the white conservatives' site, and both of the KKK sites.

Racial separatism was another common theme to emerge from the sampled hate sites. Not only was racial separatism a common topic on both black separatist sites, it also emerged as a major theme of the Christian Identity Web site and both of the KKK sites. However, racial separatism was never the main theme of any of these five sites, even the two black separatist ones.

Of particular relevance to this study was the extensive amount of political and religious speech (both of which are grounded in the First Amendment guarantees of free expression) that was present on the sampled sites. Political speech was a major component of all the sampled hate sites, except for the Christian Identity site, which was devoid of politically oriented content. Indeed, political speech was the main element of the sites operated by the neo-Confederate group, the white conservative organization, and one of the black separatist groups (NBPP). Religious speech was also a major component of the vast majority of the sampled hate sites, excepting the racist skinhead site, where it was completely absent, and the sites operated by the neo-Nazi group and one of the black separatist groups (NBPP), where it was only a minor part. In fact, religion was the main theme of nearly half of the sampled sites, including the Christian Identity site, one of the black separatist sites (NOI), and both of the KKK Web sites.

Moreover, as indicated in this summary, half of the sampled sites included extensive amounts of both political and religious content. Included in this category were the neo-Confederate site, the white conservatives' site, one of the black separatist sites (NOI), and both of the KKK sites.

However, some of the themes to emerge from the sampled hate sites were never anything more than minor ones. The white conservatives' site had a minor theme about preserving Southern heritage. A KKK site (MWKKKK) contained minor themes related to preserving the Mississippi state flag, as well as Western culture and civilization. Additionally, the racist skinhead site, the neo-Confederate Web site, and one of the KKK sites (AWKKKK) had what might be called a minor "social commentary" theme.

# Chapter 8

# Targets of Hatred on Internet Hate Sites

Hatred is a theme that appeared in the rhetorical discourse found on all of the sampled hate Web sites that was not specifically identified because it is implicit in their being called "hate" sites. While online hates sites contain a great deal of hateful commentary regarding nations, organizations, and governments, many of the messages that U.S.-based hate groups post on the Internet express a hatred for persons or groups of persons. This chapter identifies what persons were the targets of hatred on the publicly accessible areas of the 10 U.S.-based hate sites examined in this study.

## Ku Klux Klan

**American White Knights of the Ku Klux Klan (AWKKKK)**

AWKKKK directed hatred toward a multitude of persons on its site, thus belying its contention that it is not a hate group. As for specific individuals, AWKKKK made hateful comments about O. J. Simpson, Morris Dees and other members of the SPLC, and in particular, Martha Burk, the person responsible for spearheading protests against the Augusta National Golf Club for its policy of denying membership to women.

Most of AWKKKK's hatred, however, was directed toward groups of persons, such as blacks, Mexicans, Hispanics, Indians, Asians, and biracial persons, who the group derogatorily referred to as "mongrels." Like most Klan organizations, AWKKKK harbors an extreme dislike for nonwhites, a hatred the group expressed throughout its Web site. As an example, AWKKKK disparaged nonwhites by referring to them as "muds." Of all the nonwhites that the group mentioned online, AWKKKK directed the majority of its racially based hatred against blacks. In addition to using all sorts of derogatory terms to refer to blacks, AWKKKK made several disparaging comments about members of the black race. However, AWKKKK is an equal opportunity hate group, as evidenced by the fact that the organization also expressed hatred toward some whites on its site. For example, AWKKKK indicated that it hates whites who blame their problems on Jews and nonwhites instead of evaluating their own conduct. Ironically, however, AWKKKK seemed to blame their problems on Jews and nonwhites on its Web site.

In accordance with its religious beliefs, AWKKKK denounced homosexuals, who the group regards as being immoral. As it contends that Christianity is superior to all other religions, AWKKKK also directed hatred against non-Christians, namely Jews and Muslims. Additionally, AWKKKK expressed an animosity toward child abusers and drug users, persons that the group refuses to permit as members. AWKKKK's hatred for child abusers led it to provide links to sites maintained by groups like the National Center for Missing and Exploited Children.

However, an examination of AWKKKK's hate pages revealed that Jews are overwhelmingly the targets of the group's hatred. At various places on its site, AWKKKK claimed Jews control virtually all aspects of American society. For one, AWKKKK alleged that due to the large number of Jewish-owned media in the United States, Jews have been able to manipulate American public opinion in their favor. The group claimed, for example, that Jewish-owned media in the United States often manipulate American public opinion by providing false information. AWKKKK also expressed its belief that Jews have assumed control of the various levels of U.S. government and have wielded this governmental power to transform American society into something resembling a Jewish state. As an example, the group accused Jews of trying to replace Christian symbols, such as the cross, with Jewish icons, like the Star of David. Moreover, AWKKKK alleged that Jews have utilized their supposed control of the U.S. government to initiate and support American military operations against various Middle Eastern countries that have had strained relations with Israel, and consequently have made Israel's enemies turn their sights on America. The group argued that this allegation is clearly supported by the events of 9/11.

As part of its hatred for Jews, AWKKKK also expressed hostility toward Israelis. For example, the group praised attacks against Israeli Jews, such as those perpetrated by Palestinian suicide bombers. In addition to its own anti-Jew rhetoric, AWKKKK provided links to a variety of externally maintained anti-Semitic sites, like www.removejews.com.

**Mississippi White Knights of the Ku Klux Klan (MWKKKK)**

MWKKKK's views on religion, racial segregation, and the supremacy of the white race form a combustible mix that fuels the organization's hatred of several persons. On its site, MWKKKK expressed hatred for a variety of groups, including Mexicans, Asians, blacks, Jews, homosexuals, bisexuals, and biracial individuals. MWKKKK's animosity toward Mexicans and Asians was implicit in its online application which stated that the organization denies membership to individuals belonging to these groups. However, MWKKKK's hatred for the remaining persons was more explicit.

Although MWKKKK steadfastly maintains that it does not hate blacks, this contention was belied by the content on its site.

MWKKKK repeatedly used the term "Negroes," which many in the black community consider offensive, and made a variety of disparaging comments about blacks, like suggesting that they are somehow mentally inferior to whites. In addition to joining AWKKKK in derogatorily referring to biracial individuals as "mongrels," MWKKKK dedicated an entire page to discussing why it believes whites should not produce biracial offspring. For instance, the group claimed that whites pollute the white gene pool when they have biracial children. However, MWKKKK only discussed how it believes blacks' genes will pollute the white gene pool, a fact which further belied the group's contention that it does not hate the black race. Whatever ambiguities may exist with regard to MWKKKK's feelings toward blacks, the group made no attempt to conceal its hatred for Jews on its Web site. Not only did MWKKKK openly refer to Jews as "vile people," the group persistently ridiculed the Talmud and provided links to hate sites operated by anti-Jewish groups. MWKKKK attempted to justify its ill will toward Jewish people by citing scripture, including that found in the Talmud. The group also cited scripture when attempting to defend its hatred of homosexuals, bisexuals, and biracial individuals.

However, child molesters were the persons to which MWKKKK expressed the most hatred on its Web site. As indicated earlier, the group's site had several hate pages dealing with these sexual predators. Not only did MWKKKK express its intense dislike for child molesters, the group made every effort to help ensure that these individuals were subjected to constant public scrutiny. Along with petitioning Mississippi's government to mandate more extensive community notification regarding the whereabouts of sexual offenders, MWKKKK assisted persons in locating child molesters in the Mississippi area by providing a link to the state's sex offender registry. Sometimes the group even listed the names of known offenders living in Mississippi. In one instance, MWKKKK took credit for raising public awareness of, and convincing a local television station to do a story about, a convicted sexual predator who the group claimed was attempting to lure children into his home.

## Neo-Nazi

### National Socialist Movement (NSM)

Inspired by Hitler, NSM directed hatred toward a variety of persons on its Web site. As part of its belief in the supremacy of the white race, NSM expressed a hatred for all nonwhites, which, like AWKKKK, the group disparagingly referred to as "muds." Not only did NSM regard all nonwhites as being inferior to whites, the group suggested that they should not be allowed to migrate to, or remain living in, the United States. Indeed, the group went so far as to state that all nonwhites should be murdered at one point on its site. While most of NSM's hatred for nonwhites was directed at blacks living in the United States, the group also made disparaging comments about people of color from Mexico, China, India, Sri Lanka, and all of Asia and Africa. However, regardless of what group of nonwhites it was discussing on its site, NSM almost always used derogatory terminology when referring to them. Additionally, NSM posted a variety of racist jokes and comics and provided links to several other Web sites that exhibited hatred for the various nonwhite groups it denigrated on its site. The group expressed its hatred for whites that mix with nonwhites as well.

NSM also expressed an intense dislike for Arabs, drug dealers, homosexuals, and child molesters. The group accused Arabs of being torturers, cold-blooded killers, and repressors of women. NSM explained that it hated drug dealers because it firmly believes that drugs not only harm individuals, but society as a whole. In addition to regarding homosexuals as "unnatural freaks" who should not be allowed to live in the United States, the group posted several antigay jokes and comics on its site and provided links to a variety of antihomosexual Web sites. The hatred that NSM expressed for child molesters was even more intense. The group stated that child molesters, like nonwhites, should be killed.

Extremely pro-Christian, NSM also expressed animosity toward Muslims and Jews. At various points on its site, the group indicated its belief that Muslims pose a threat to the security of various Western countries, such as Great Britain. Like AWKKKK and MWKKKK, however, NSM directed most of its hatred against Jewish persons, in particular Israeli and American Jews. Along with posting a variety of anti-Semitic jokes and comics and using a host of derogatory terms when referring to Jews, NSM provided a list of reasons to justify its hatred for the Jewish people. The group stated that it believes Jews are generally dishonest and murderous persons who work tirelessly to destroy white Christians around the world, and in the United States in particular. With regard to its allegations of Jewish dishonesty, NSM claimed that Jews invented the Holocaust as a means of generating sympathy and stifling legitimate criticisms of their actions. The group also made several allegations in an attempt to support its assertion that Jews are a generally murderous people, such as claiming that Jewish persons helped cause, had prior knowledge of, and were pleased by 9/11.

Because of its beliefs, NSM explained that it feels that Jews, like nonwhites and homosexuals, should not be allowed to live in the United States. As it did with respect to nonwhites, the group even went so far as to state that all Jews should be killed at one point on its site. NSM supplemented its own anti-Jew rhetoric by posting anti-Semitic comments that were supposedly made by various historical figures and well-known persons, such as Hitler and Benjamin Franklin. In addition, the group provided links to pro-Nazi, anti-Jew, anti-Israel, and Holocaust denial Web sites.

NSM also expressed hatred toward several specific individuals on its site. As part of its never-ending anti-Semitic assault, NSM disparaged particular Jews living around the world, such as Israel's Ariel Sharon, who the group referred to as a "fat kike" and a "bloated tick of a Jew." NSM also indicated its intense dislike for some members of the Anti-Defamation League (ADL), which it accused of being a racist, anti-American hate group that spies on Americans. In addition, NSM directed hateful comments toward some non-Jewish government officials, in the United States and abroad, who it claimed had "sold out to the Jews," such as George W. Bush, Dick Cheney, Colin Powell, Arnold Schwarzenegger, and Tony Blair. What is more, the group demonstrated an animosity toward specific members of the media that it accused of providing slanted news coverage in favor of Israel and Jews, like CNN's Wolf Blitzer.

## Racist Skinhead

### Tualatin Valley Skins (TVS)

One would be hard-pressed to come up with an organization that more accurately fits the definition of a hate group than TVS. TVS lashed out against a multitude of persons and used a seemingly endless array of derogatory terms (e.g., "mud people," "subhumans") to refer to them on its hate site.

Reflecting its contempt for law and order, TVS directed hateful comments toward lawyers and police. A raging antihomosexual group, TVS disparagingly referred to homosexuals as "faggots" and even made light of the fact that AIDS has had a devastating effect on the homosexual community. Avowed white supremacists, TVS expressed hatred toward all nonwhites, especially blacks. The group made a host of racist remarks about blacks. TVS's hatred also extended to biracial individuals and whites that they regarded as race traitors. In addition, TVS directed hate-laden rhetoric at specific groups of foreigners, in particular Asians, Africans, Indians, Israelis, Ethiopians, Pakistanis, Puerto Ricans, and Mexicans, who it repeatedly disparaged.

The hatred that TVS expressed for Israelis was, however, just one part of the group's larger anti-Jewish sentiment. Along with denouncing Judaism, TVS applauded the atrocities that Hitler committed against Jews. The group also accused Jews of controlling the media, government, entertainment industry, and educational system in the United States and the world, and using that power to brainwash American minds. TVS's anti-Semitism was reflected on virtually every one of its hate pages.

TVS also indicated its hatred for several specific individuals, including Presidents Bill Clinton and George W. Bush. TVS referred to President Clinton as a "nigger" and regarded President Bush as an idiot, a claim that the group attempted to substantiate by posting some of Bush's own statements. Additionally, TVS made hateful remarks about several televangelists (e.g., Falwell, Graham), and conservative commentators of the media (e.g., Limbaugh, Hannity).

## CHRISTIAN IDENTITY

### Orange County Assembly of Christ (OCAC)

Not content with simply acknowledging its approval for some white supremacist groups, OCAC joined them in directing hatred toward various persons, namely homosexuals and Jews. OCAC regarded homosexuality as an abomination and expressed its hatred for all persons who engage in homosexual acts. In fact, one of OCAC's hate pages was a promotion for a protest that the church was planning to hold at a religious television network which employed a man that had allegedly engaged in homosexual acts with various males, including his own son.

However, OCAC's hatred for homosexuals paled in comparison to the ill will the church expressed for Jews on its site. OCAC not only implied that it was anti-Jew by devoting a substantial portion of its site to arguing that whites, not Jews, are the "true Israelites," but also the church's anti-Semitism was clearly reflected in statements like "the Jew is a parasite." Furthermore, in soliciting persons to join it in protesting against the alleged homosexual previously mentioned, who perhaps not coincidentally was also supposedly Jewish, the group made a point of stating that no Jews were allowed. OCAC defended its anti-Semitism by pointing to verses from the Christian Bible and the expulsion of Jews from various countries throughout history, something which the church attempted to document on one of its hate pages. OCAC's anti-Semitic rhetoric directly contradicted its claim that the group promoted peace.

## Black Separatist

### Nation of Islam (NOI)

In addition to the animosity which NOI exhibited toward the U.S. government in its online political rhetoric, the group also expressed a hatred for some members of the media on its site. In particular, the organization expressed hatred for those members of the media that it claimed had mischaracterized its organization, and its leader Louis Farrakhan, as being antiwhite, anti-Semitic, antigay, anti-Christian, anti-Catholic, and anti-American. The group also indicated that it believed various members of the media had unjustly vilified the Islamic religion, Farrakhan, and other NOI members.

Even though it is understandable why one might characterize NOI as being anti-American, if the organization's online rhetoric at the time of this study was any indicator, the group had every reason to claim that the other aforementioned characterizations of its organization were unjustified. As far as being antigay, homosexuality was not even mentioned on NOI's site. NOI's site also did not contain any antireligion rhetoric. In fact, the group repeatedly indicated that it was seeking to promote mutual understanding and peaceful coexistence between all people who believe in God, regardless of their religious denomination. Although it is opposed to interracial marriage and race-mixing, NOI did not express any hostility toward people of other races or national origins on its site. In fact, the group stated on one of its pages that people of every race and nation are God's people.

As for being anti-Semitic and antiwhite, NOI did not make any statements on its Web site that suggested it had any ill will toward Jews and whites. The group was, however, critical of Jews and whites. For example, while NOI accused both Jews and whites of treating blacks poorly, it maintained that they were not the black community's real problem. The group stated that blacks must serve their own needs and take personal responsibility for whatever situations befall them. Simply put, whatever reasons there may be for claiming that NOI disseminates anti-Semitic and antiwhite hate speech in the offline world, there was nothing in the organization's online rhetoric at the time of this study to support such a claim.

**New Black Panther Party (NBPP)**

Consumed by feelings of discrimination and victimization, NBPP directed hatred against several persons on its hate site. While claiming not to be racist or anti-Semitic, the organization expressed hatred toward both whites and Jews generally, as well as specific individuals belonging to these two groups. For example, along with all of the other hateful rhetoric that it directed at whites in the United States and throughout the world, NBPP derogatorily referred to white persons as "crackers" and "rednecks." NBPP even had a section on its site dedicated to particular whites that the group hated, such as Fox News' Bill O'Reilly, the late Rev. Jerry Falwell, and former Homeland Security Director Tom Ridge. NBPP's site also included some anti-Semitic comments, such as a reference to Jews as "bloodsuckers." As it did with whites, the group also expressed hatred toward particular Jews, such as Israel's Ariel Sharon. Like NSM, NBPP also indicated its intense dislike for some members of the ADL, which it accused of being a racist, criminal organization.

In addition to its hateful rhetoric regarding whites and Jews, NBPP's reparations rhetoric seemed to suggest that the group bears ill will toward Arabs on account of their ancestors' roles in the African slave trade. However, like some of the other groups in this study, NBPP is an equal opportunity hate group. As such, NBPP made hateful comments about several specific blacks who it regarded as black race traitors, ranging from scholars to media figures (e.g., Oprah Winfrey). Like TVS, NBPP also disparaged homosexuals by referring to them as "faggots."

Most of NBPP's hateful rhetoric, however, was directed at government officials. NBPP made disparaging remarks about several government officials, including Colin Powell, Condoleeza Rice, Clarence Thomas, John Ashcroft, Donald Rumsfeld, Dick Cheney, and both Bush presidents. For example, the group referred to President George H. W. Bush as "the cracker you love to hate." NBPP also expressed an animosity toward police officers, in particular those working for the Chicago, Cincinnati, and New York police departments, departments which have gained a reputation (deservedly or undeservedly) for brutalizing and murdering blacks. NBPP even directed its hatred at specific police officers, both white and nonwhite. As an example, the group used the term "murderous pigs" to refer to six Cincinnati police officers, all of which were identified by name and photograph on its site, charged in connection with the murder of a black man.

## Neo-Confederate

### League of the South (LOS)

LOS's pro-South sentiment was reflected in what persons the organization expressed hatred for on its site. Most of LOS's hateful rhetoric was directed against those persons who the group regarded as being anti-South, such as Rev. Jesse Jackson, liberals, Northerners, members of the NAACP, and individuals lobbying for the removal of Southern symbols (e.g., Confederate battle emblem) from public properties in the South. Indeed, Northerners and NAACP members were among the most frequent targets of LOS's hatred. LOS's anti-Northerner rhetoric included claims that, in general, Northerners are morally bankrupt and less hospitable than Southerners. As for members of the NAACP, LOS regarded them as the "Klan with a tan," on account of the numerous public demonstrations that they participate in throughout the Southern United States. The group also expressed a hatred for individuals it regarded as traitors against the South, such as particular Southern politicians who supported the removal of the Confederate battle emblem from Southern states' flags.

LOS expressed ill will toward several persons that were also the targets of hatred on some of the other hate sites in this study.

LOS expressed an animosity toward several members of the media, like Fox News reporter Geraldo Rivera. The group also exhibited an anti-immigrant sentiment on several of its hate pages. LOS indicated that its anti-immigrant attitude grew primarily out of its belief that foreign immigrants, both white and nonwhite, typically want to transform the prevailing U.S. "Southern culture" into a culture more reminiscent of what they were accustomed to in their native lands. Because of 9/11, LOS also exhibited an animosity toward Muslims. For example, LOS stated that the only peaceful Muslims are those that do not have a strong commitment to Islam, which the group called an essentially violent religion. In addition, LOS expressed hatred for Morris Dees and other members of the SPLC who it claimed had unfairly labeled the organization as a racist hate group, something which was not evident from the group's online rhetoric. Although it expressed animosity toward Jesse Jackson and members of the NAACP, LOS's hatred for these persons did not appear to be based on race.

## OTHER HATE GROUPS

### Council of Conservative Citizens (CCC)

While CCC claims that it does nothing more than advocate the rights of conservative white European-Americans, the organization clearly expressed an animosity toward several persons on its site. A group with both strong prowhite and anti-immigrant views, CCC spoke ill of nonwhite illegal immigrants that had entered the United States. For example, the group suggested that nonwhite illegal immigrants put white European-Americans at a greater risk of contracting diseases, becoming a victim of crime and terrorism, as well as experiencing depressed wages, loss of jobs, and a declining standard of living.

However, CCC also expressed hatred for nonwhites legally residing in the United States. The group claimed that people of color, in particular blacks and Hispanics, have burdened white European-Americans with more crime and the responsibility of maintaining welfare, affirmative action, multicultural diversity programs, and so forth. CCC exhibited a particular disdain for blacks. In addition to providing links to several antiblack sites, the group claimed that blacks are mentally inferior to whites and that black males are culturally and genetically more prone to violence than men in any other racial group.

As part of its Christian belief system, CCC regarded homosexuals as moral degenerates on its site. CCC even expressed hatred toward obese welfare recipients, who the group accused of getting fat off of the taxpayer's plate. However, even here the discussion only focused on Hispanic and black welfare recipients who were overweight, or as the organization disparagingly referred to them as "brown and black jelly-bellies." An anti-Jewish sentiment also ran through CCC's hate site. Not only did CCC suggest that certain Jews have tried to promote a black-white race war, the group also provided links to sites with anti-Semitic themes. The existence of these links, coupled with the racist comments on the site, contradicted CCC's claim that it is not a hate group.

**Jewish Defense League (JDL)**

Simply put, JDL expressed hatred toward any person or persons that it regarded as being anti-Jew. This included Klansmen, neo-Nazis, Holocaust deniers, Christians that try to convert Jews to Christianity, specific government officials, certain private U.S. citizens, and public figures, such as director Michael Moore and political pundit Pat Buchanan. JDL also made hateful comments about Jewish persons who did not stand up for their fellow Jews, like during the Holocaust. The group even directed hatred toward O. J. Simpson, accusing him of killing Nicole Brown Simpson and Ronald Goldman, a Jew.

However, the majority of JDL's hatred was directed at Muslims and Arabs. Muslims were repeatedly regarded as terrorists on the group's site. As for Arabs, they were referred to as a "malignant" and "murderous" people who were the enemy of Jews. The anti-Arab rhetoric on JDL's site was primarily focused on Palestinians and Saudi Arabians. JDL expressed hatred for several Palestinians, including Palestinian suicide bombers, as well as members of Hamas and the Palestinian Liberation Organization (PLO). Interestingly enough, JDL accused Saudi Arabians of trying to take over the world, the very same accusation that several of the other hate groups in this study made about Jews.

## SUMMARY

This chapter revealed that a vast array of persons were the targets of hatred on the 10 U.S.-based hate sites selected for inclusion in this study. While some persons emerged as objects of hatred on several hate Web sites, others only had animosity directed against them on a particular site.

Excepting the Christian Identity site and the white conservatives' site, all of the sampled hate sites included some rhetoric expressing hatred toward particular individuals. Several of the sampled hate sites contained hateful comments directed at specific Jews (NBPP, NSM), private U.S. citizens (JDL, MWKKKK), public figures in America (AWKKKK, JDL, LOS), government officials (JDL, LOS, NBPP, NSM, TVS), members of the media (JDL, LOS, NOI, NSM, TVS), and members of antihate organizations (AWKKKK, LOS, NBPP, NSM), such as the ADL and the SPLC. In addition, certain televangelists were the objects of hatred on the racist skinhead site, while one of the black separatist sites (NBPP) included some disparaging remarks about specific whites, police officers, and blacks who were regarded as black "race traitors."

While particular individuals were the targets of hatred on some Web sites, the majority of the hateful person-centered rhetoric on the sampled sites was directed at groups of persons, typically very large groups. Of all the groups, however, Jews were undoubtedly the group of persons most frequently attacked. Not only was anti-Jew rhetoric found on all but three of the sampled Web sites (JDL, LOS, NOI), it often represented the majority of the hateful content on those sites. Hateful antihomosexual rhetoric also appeared on all but three of the hate sites included in this study (JDL, LOS, NOI), but it generally paled in comparison to the anti-Jew rhetoric both in terms of its volume and viciousness.

After Jews and homosexuals, the most common targets of hatred on the sampled sites were nonwhites. This was not all that unexpected given that most of the hate sites in this study were maintained by whites. Hateful comments regarding nonwhites generally, and blacks specifically, appeared on half of the sampled sites (AWKKKK, CCC, MWKKKK, NSM, TVS). While most of those comments were directed at nonwhites living in the United States, the neo-Nazi site included derogatory remarks about people of color living in Mexico, China, India, Sri Lanka, and all of Asia and Africa.

Other groups of persons that were objects of hatred on several of the sampled sites included: child molesters (MWKKKK, NSM); biracial individuals (AWKKKK, MWKKKK, NSM, TVS); Mexicans (AWKKKK, MWKKKK, NSM, TVS); Asians (AWKKKK, MWKKKK, TVS); Hispanics (AWKKKK, CCC); Indians (AWKKKK, NSM, TVS); Israelis (AWKKKK, NSM, TVS); Arabs (JDL, NBPP, NSM); Muslims (AWKKKK, JDL, LOS, NSM); Palestinians (JDL, TVS); police officers (NBPP, TVS); and foreign immigrants to the United States (CCC, LOS). However, some groups of persons were targets of hatred on only one of the sampled hate sites: child abusers (AWKKKK), drug users (AWKKKK), drug dealers (NSM), lawyers (TVS), whites (NBPP), liberals (LOS), Northerners (LOS), NAACP members (LOS), Klansmen (JDL), neo-Nazis (JDL), Holocaust deniers (JDL), obese welfare recipients (CCC), Africans (TVS), Chinese (NSM), Sri Lankans (NSM), Saudi Arabians (JDL), Ethiopians (TVS), Puerto Ricans (TVS), whites that mix with nonwhites (NSM), and Christians that try to convert Jews to Christianity (JDL). In addition, the racist skinhead group exhibited an extreme dislike for individuals it called "white race traitors," the neo-Confederate group demonstrated an animosity toward people it dubbed "traitors to the South," and the Jewish organization expressed a hatred for a class of persons it termed "Jewish traitors."

That the targets of hatred on the sampled hate sites tended to be very large groups of persons is of significance to this study for two reasons. First, the Supreme Court's "true threats" test, one of the First Amendment tests used to evaluate hate site content in this study, is not applicable to a threat made toward a very large group of persons, especially one as ill-defined as, say, "liberals." Second, it suggests that had the constitutionality of the hateful rhetoric on the sampled sites been evaluated on the basis of its allegedly defamatory nature, an approach to evaluating hate speech supported by some, that the content would still have been found to be constitutional. As previously indicated, allegedly libelous remarks made about large groups of people sharing an innate human characteristic (e.g., race, sexual orientation) are entitled to significantly more, if not absolute, First Amendment protection than are defamatory comments made about individuals or sufficiently small groups (Overbeck, 2002).

This analysis is also important because it reveals that the hateful rhetoric posted on the sampled sites was not confined to "historically oppressed" groups in the United States. As noted earlier, some legal scholars supporting the regulation of hate speech have tended to argue that such expression is only entitled to extremely limited, if any, First Amendment protection when it includes demeaning commentary directed toward "historically oppressed" groups in America, namely blacks and Jews. While blacks and Jews were among the three most common targets of hatred on the sampled Web sites, many other groups of persons had hatred directed against them on the sites. On the sampled hate sites, homosexuals were more often the targets of hatred than blacks. Thus, this analysis did not support an argument that hate sites should be banned solely on the grounds that they contain content demeaning to "historically oppressed" groups in the United States. Instead, this analysis provided support for the use of conventional court-formulated First Amendment tests for evaluating hate site content, such as those being used in this study.

# Chapter 9

# Nontextual Content Posted on Internet Hate Sites

Twelve years after the Supreme Court first established that the First Amendment's language regarding speech and press freedoms should not be interpreted literally, it had to determine whether the amendment implicitly encompassed symbolic expression in *Stromberg v. California* (1931). *Stromberg* involved a teenage woman's felony conviction for flying a Soviet flag in violation of a California statute that prohibited the public display of, among other items, "a sign, symbol, or emblem of opposition to organized government" (*Stromberg v. California*, 1931, p. 363). In setting aside the defendant's conviction under this California statute, the *Stromberg* Court established that the right of free speech, embodied in the First Amendment and made applicable to the states by the Fourteenth Amendment, also extended to symbolic expression. Since *Stromberg*, the Supreme Court has held that the right of free speech protects music, the alteration and burning of the American flag, students in wearing black armbands on public school grounds to protest a war, and various other forms of symbolic expression or expressive conduct.

However, like the freedom to express oneself through the printed and spoken word, the freedom to engage in symbolic expression can also be constrained. The Supreme Court made this point clear in *United States v. O'Brien* (1968). In *O'Brien*, the Court held that the right of free expression did not protect antiwar protester David Paul O'Brien from being prosecuted for illegally burning his military draft card to symbolically express his opposition to the Vietnam War. In rendering its decision, the Court specifically rejected O'Brien's argument that his act of burning his draft card was symbolic speech protected under the First Amendment. Chief Justice Warren, who delivered the Court's opinion, stated, "We cannot accept the view that an apparently limitless variety of conduct can be labeled 'speech' whenever the person engaging in the conduct intends thereby to express an idea" (*United States v. O'Brien*, 1968, p. 376).

Thus, the constitutionality of a particular hate site may hinge as much on its symbolic, nontextual content as on its textual rhetoric. Accordingly, this chapter identifies the forms of nontextual content that were posted on the publicly accessible areas of the 10 U.S.-based hate sites selected for examination in this study.

## NONTEXTUAL INTERNET CONTENT

On the Internet, nontextual content can include audio, motion video, still images, or computer-generated animation. What follows is an explanation of how these various types of nontextual content were defined for purposes of this study.

The term "audio" is used when referring to any audio-based content (e.g., music, sound effects, human voice) that was present on a hate site. The term "motion video" is used to refer to video of persons, animals, or objects in a state of motion. Oftentimes motion video is recorded with a camcorder and then posted online. By contrast, the term "computer-generated animation" is only used when referring to a computer program that showed persons, animals, or objects—including text—in motion. The designation "still image" is used to refer to any stationary, nontextual visual imagery. However, this is a very general designation as various sorts of nontextual still images can be posted online. Thus, more specific terminology is used when referring to those still images that appeared on the sampled hate sites.

Still images posted on the Internet can essentially be divided into two categories: (1) "photographic images"—images that existed offline, apart from any computers, and were reproduced in a digital format suitable for posting online; and (2) "computer-generated images"—images posted online that were entirely created by computer. Virtually any image that exists in the offline world can be photographed, transformed into a digital file, and posted online. Indeed, digital cameras allow persons to photograph offline images while simultaneously saving them in a digital format that can be posted online. Computer scanners also permit individuals to transform offline images (e.g., book covers, portraits) into a format that can be posted on the Internet. Some computer programs even allow users to create digital "freeze frames" of offline images that appear on motion video.

Just as the Internet has allowed persons to engage in communication unlike that in the offline world, so too has the computer enabled individuals to create images not of this world. Thanks to computer imaging programs like Adobe Photoshop, persons can generate graphic-style (i.e., nontypeset) lettering, various forms of artwork, and any number of other computer-generated images. Generally speaking though, it requires more time and greater computer skills to generate an original image using an imaging program than to simply digitize a copy of a preexisting offline image using a digital camera or computer scanner. However, as imaging programs have become more user-friendly it has become increasingly easy for persons to design a Web site that contains computer-generated images, such as the rectangular-shaped banners that often appear at the top of Web pages. In fact, imaging programs have even made it possible for "Web masters" to alter digitized offline images or combine them with computer-generated images. These capabilities have led to the creation of a whole separate genre of images that is best termed "natural-computer image hybrids." Like anything posted online, however, computer-generated images appearing on hate pages may have simply been downloaded from another site on the Internet.

The process of differentiating between "photographic images" and "computer-generated images" (CGIs) is inevitably an inexact science, as it is often impossible to know the true origins of any still image posted online given that computer imaging programs have become fairly sophisticated. Thus, the still images that were posted on the sampled Web sites were categorized strictly based on their appearance. Still images that appeared to have existed in the offline world were classified as "photographic," and all other images, including natural-computer image hybrids, were placed into the "computer-generated" category. Regardless of how still images were categorized, the determination of whether the sampled Web sites contained constitutionally proscribable content, the main focus of this study, was in no way compromised. After all, an online image's constitutionality is not dependent on whether it originated from the offline world or was generated on a computer.

## KU KLUX KLAN

### American White Knights of the Ku Klux Klan (AWKKKK)

Like most Web sites, AWKKKK's hate site contained an extensive number of still images, mainly computer-generated ones. Indeed, AWKKKK's home page contained nothing more than a CGI upon which the group's name appeared in graphic-style lettering. The evidence suggested that AWKKKK changes this image periodically as it was a Christmas scene on December 1, 2004. The group's site also contained several CGIs that were natural-computer image hybrids, like digitally altered photographs (e.g., woman's photo modified to show horns jutting out from her head) and offline photos (e.g., an ape, black woman) that had been combined with images created on a computer. Additional CGIs on AWKKKK's site included flags (e.g., Confederate), graphs, artwork (e.g., Halloween-type), people (e.g., adults, children, Indians, Klansmen), animals (e.g., monkeys, lions, dinosaurs, rats), various objects (e.g., clothes, gavel, monument, detergent box), hate symbols (e.g., swastikas), corporate trademarks, and other hate groups' logos. AWKKKK claimed that some of the CGIs it had posted on its site were "courtesy of" an Athens, Georgia television station's Web site. Several of the CGIs on AWKKKK's site contained hateful (e.g., racist, anti-Semitic) or religiously oriented comments (e.g., scripture, morals).

Sometimes the CGIs on AWKKKK's site simply served as a visual supplement to the textual rhetoric posted on a Web page, including that contained on printer-friendly, PDF-formatted Klan flyers that the group had posted online. However, more often than not the CGIs on the group's site were utilized as eye-catching "hyperlinks" (i.e., nontextual links to Web pages). For example, AWKKKK's site contained a profuse amount of computer-generated, banner-style hyperlinks and even had several pages that contained almost nothing but these hyperlinks. While most of the CGIs that functioned as hyperlinks on AWKKKK's site simply connected persons to the group's other Web pages, many served as links to Web sites maintained by other like-minded Klan organizations. Other CGIs on AWKKKK's site served as hyperlinks to Web sites operated by government (e.g., Congress), news organizations (e.g., Fox News), and groups selling everything from pendants to prison art, and with ideological underpinnings ranging from antiblack and anti-Semitic to profamily, prochildren, prowhite, and pro-Christian.

AWKKKK's hate site also contained numerous photographic images, many of which were photos of the group's public demonstrations. The demonstration photographs on AWKKKK's site included images of flags (e.g., American, Georgia), crowds, individuals (e.g., police, onlookers), police cars, persons delivering speeches, the Ten Commandments, and articles (e.g., banners) that the group displayed while demonstrating. Some of these demonstration photographs were also obtained from an Athens, Georgia television station's Web site.

AWKKKK also posted photos of offline activities engaged in by other hate groups, both KKK (i.e., Cleveland Knights of the Ku Klux Klan) and neo-Nazi (i.e., NSM), including meetings, as well as cross and swastika lighting ceremonies. The meeting photos showed Klansman and neo-Nazis, both of which were wearing their respective group's outfits (i.e., robes, Nazi regalia), listening to, and delivering, speeches. In many of these photos, Klansmen and neo-Nazis could be seen raising their hands in a Nazi salute. As for the lighting ceremony photos, they consisted of images of inanimate objects (e.g., sticks) and persons, both children and adults, isolated and congregating with others at an outdoor facility during daytime and nighttime when the ceremonies were taking place. AWKKKK's site also included photographic images of babies, animals, buildings, magazine and CD covers, various objects, and people like Pat Buchanan, individual Africans, a group of armed U.S. military personnel, and a man in military fatigues standing in a room in front of a rifle. AWKKKK incorporated comments, some racist, into several of the photos it posted online. Some of the photos on the group's site also served as hyperlinks to external Web sites.

AWKKKK also utilized computer-generated animation, mainly in its "Kristian Kids Korner," where there were several video games that could be played and two cartoon-like animated sequences that could be viewed. While most of the video games in the "Kristian Kids Korner" were unsophisticated by today's technological standards, the two computer-generated animated sequences found in that section were a bit more technologically advanced. One computer-generated animated sequence showed a dinosaur and graphic-style text in motion. The other sequence was intended to symbolically represent the group's belief that persons wash away their sins when they support the Christian faith. That sequence had a sweater, with text (e.g., "sins," "spiritually dirty") superimposed on it, and a Tide-like logo and detergent box, both inscribed with the name "Jesus" and other religious terminology, moving about the screen. AWKKKK's site also contained some computer-generated animation of fire, which sometimes emanated from objects or engulfed words.

Audio was also a component of AWKKKK's site. Visitors to the "Kristian Kids Korner" had the option of listening to lyric-free music and other sounds (e.g., a dinosaur's roar) while viewing the computer-generated animated sequences in that section. Persons visiting the other areas of the group's site were exposed to lyric-free, Southern-style music (e.g., country). All of the audio on AWKKKK's site emanated from a player which users were able to turn off.

**Mississippi White Knights of the Ku Klux Klan (MWKKKK)**
Still images constituted nearly half of the overall content found on MWKKKK's hate Web site, and the majority of these images were photographs. There were photographs of rallies held by MWKKKK itself and by other KKK groups during the 20th century, like the Klan's various marches in Washington D.C. in the 1920s. These rally photos showed KKK memorabilia (e.g., hooded dolls) and other artifacts (e.g., American and Confederate flags), as well as persons, both hooded and unhooded, engaging in acts such as holding guns, burning crosses, delivering speeches, parading down streets, and simply gathering together. On the group's merchandise page there were photographs of various items, like flags (e.g., Confederate, Mississippi), quilts, Klan member dolls, and portraits of Nathan Bedford Forrest. As part of its campaign against sexual predators, MWKKKK also posted photos of convicted child molesters. Additional photographic images on the group's site included those of former Klan members and leaders, old KKK flyers and posters, and several other Klan artifacts, such as a poster for D. W. Griffith's pro-Klan movie *The Birth of a Nation*.

As for CGIs posted on MWKKKK's site, most were of the organization's logo, which contains the group's name and initials, as well as a sword, feather, and the Klan "blood drop" (i.e., a white cross with a blood drop in its center) on a red circular background. MWKKKK also posted CGIs of flyers used by the group to advertise its rallies and to honor its fallen members. Other CGIs on MWKKKK's site consisted of crosses, animals, various flags (e.g., Mississippi, Confederate, MWKKKK), an outline of Mississippi, the Klan "blood drop," banners for other hate groups' Web sites, and images of individual Klan members riding horses and holding various objects (e.g., torch, Mississippi flag). Some of the CGIs on

MWKKKK's Web site served as hyperlinks to hate sites operated by neo-Nazi organizations (e.g., White Revolution) or other Klan groups (e.g., Orion Knights of the KKK, Texas Knights of the KKK, Louisiana White Knights of the KKK).

Although still images were the predominant form of nontextual content used by MWKKKK, there was also some basic computer-generated animation on its site. Appearing on the group's home page was a simple animated sequence where a page peeled off and revealed a fully robed Klan member with a Bible in one hand and a cross in the other.

## Neo-Nazi

### National Socialist Movement (NSM)

Not only did NSM post a plethora of textual content online, the group also used an abundance of nontextual rhetoric on its site. Indeed, with the exception of motion video, NSM utilized every available nontextual method of presentation that can be employed on a Web site.

Like MWKKKK's site, the nontextual content posted on NSM's Web site consisted mainly of photographic images, many of which were of the group's public demonstrations. NSM's public demonstration photos featured its male and female members, who usually appeared wearing Nazi-style military uniforms. NSM members were shown marching, delivering speeches, giving Nazi salutes, being covered by the press, getting heckled by protesters, displaying Nazi symbols and NSM signs, or merely congregating, sometimes with other hatemongers (e.g., Klansmen). NSM also posted several photos of its members at the group's and other hate groups' private meetings and gatherings, where they were once again shown dressed in Nazi regalia and engaging in many of the same activities previously described. Occasionally, NSM members were shown burning crosses or swastikas at outdoor gatherings. The group also posted photos of its members gathering together socially and dressed in civilian attire (e.g., jeans, T-shirts), at private residences and public places, like restaurants. In some cases, NSM members were shown symbolically expressing their hatred for Israel and Jews by standing on Israel's flag. Oftentimes, however, NSM members were shown doing nothing more than posing for pictures.

NSM posted a variety of photographs of persons that it hates, like Israelis, Jews, blacks, and Mexicans. As an expression of its support for Nazism and Socialism, the group posted several photos of Hitler, both by himself and surrounded by his followers, as well as other well-known persons who endorsed these political philosophies. NSM also used photos to symbolically express its opposition to certain issues, such as posting pictures of dead fetuses to show its disapproval of abortion. However, most of the photographic images that NSM posted online were of the merchandise (e.g., books, flags, daggers, watches, jewelry, clothes, CDs) that the group had for sale, which was primarily white supremacist in nature (e.g., inscribed with swastikas). The group's site also included photos of whites (e.g., babies, adults), Nazis, buildings (e.g., mosques), objects (e.g., microphone), printed matter (e.g., articles), military personnel, government officials (e.g., President George W. Bush), law enforcement officers, Holocaust revisionists, corpses of Holocaust victims, and white individuals dressed in NSM or Nazi-style clothing. Several of the photos on NSM's site appeared in the many racist, anti-Semitic, and prowhite flyers that the group had posted online. In rare cases, NSM incorporated comments, some hateful, or computer-generated artwork (e.g., Star of David) into the photographs that it posted online.

NSM's site also contained a large number of CGIs, the most common being the group's symbol, a crest with four quadrants (i.e., black, red, white stars on blue background, red and white stripes) on which there is a swastika and the letters "NSM" inside a circle. Almost as prevalent as that CGI was a computer-generated banner, which included the group's logo and a collage of Nazi banners and symbols that appeared at the bottom of most of NSM's hate pages. CGIs of swastikas and other Nazi symbols appeared throughout NSM's site. NSM posted some other neo-Nazi groups' logos as well.

NSM's site also included a variety of computer-generated artwork depicting persons that the group hates (e.g., blacks, Mexicans, Jews,

Hispanics, Muslims, Arabs), as well as those that it admires, such as Hitler, Nazis, whites, former neo-Nazi leaders (e.g., George Lincoln Rockwell), and early American patriots (e.g., Benjamin Franklin). Much of the artwork depicting Nazis and whites on the group's site was reminiscent of the Nazi propaganda posters used during Hitler's reign. As a sign of its intense hatred for Israel and Jews, NSM posted CGIs of persons (e.g., stick figures) showing disrespect for the Jewish Star of David by, for example, throwing it into a trash can. The group also frequently used CGIs of the Star of David as a visual supplement to its text-based anti-Semitic commentary. Additional CGIs on NSM's site consisted of corporate logos (e.g., Yahoo!) and artwork depicting buildings (e.g., World Trade Center), animals (e.g., bald eagles), people, various flags (e.g., U.S., Confederate, Great Britain, POW-MIA), the Statue of Liberty, and a variety of other objects (e.g., guns, radios). As with the photos on its site, many of the CGIs used by NSM appeared in the various flyers that the group had posted online. In some cases, NSM merchandise, like a downloadable NSM computer desktop, appeared as a CGI. Several of the CGIs on NSM's site served as hyperlinks to externally maintained Web sites, such as those operated by racist "White Power" music companies.

In addition to still images, NSM used computer-generated animation. On its home page, NSM had a computer-generated animated sequence, which appeared over a backdrop of Nazi imagery (e.g., swastikas), in which text and objects moved around the screen while the group's name slowly appeared. As a further sign of its hatred for Israel and Jews, the group posted computer-generated animation of Israel's flag being torched. NSM's site also had computer-generated animation of flags waving and objects being engulfed in flames.

Audio was utilized on NSM's site as well. For one, a thunder-like sound effect could be heard while the computer-generated animated sequence was taking place on the group's home page. NSM also posted prerecorded online "radio" broadcasts, some of which included interviews with other white supremacists (e.g., Klansmen), as well as audio from its private meetings and public rallies. Each of these audio segments was approximately an hour in length and included music (e.g., "White Power"), narration, and sound effects. Taken as a whole, these audio segments were a mix of music, narration, sound effects, profanities, historical facts, self-promotion, political speech, pro-Christian commentary, pro-American discourse, prowhite rhetoric, statements regarding actual world events, and hateful comments about government, Jews, immigrants, foreigners, nonwhites, and other groups.

## Racist Skinhead

### Tualatin Valley Skins (TVS)

Like NSM's Web site, TVS's hate site included an abundance of nontextual content. This is not surprising when one considers that TVS claimed several of its members received college-level education in electronic media production. True or not, the group's site contained all forms of nontextual rhetoric.

TVS's expertise in electronic media production was reflected in the vast number of CGIs that the organization posted online, the most common of which was a rectangular-shaped banner that appeared at the top of most of the group's hate pages. That computer-generated banner included TVS's name, various words (e.g., "White Boys"), and a Jewish Star of David, encircled by a prohibited symbol (i.e., diagonal slash in a circle), superimposed over artwork of a mushroom cloud, which was supposed to represent the after-effects of a nuclear blast in Israel. As previously mentioned, TVS's Web site address was www.nukeisrael.com. TVS's site contained several other computer-generated banners combining text and artwork (e.g., swastikas) that clearly expressed the group's hatred for Jews generally, and Israelis specifically. TVS also posted computer-generated banners combining text and artwork that appeared to have been downloaded from other white supremacist groups' Web sites.

Several of the CGIs on TVS's site were natural-computer image hybrids. The group was particularly fond of altering photos of persons that it hated (e.g., blacks, Jews, Mexicans, homosexuals, particular whites) or combining them with images generated on a computer. As an example, one photo of a Jewish rabbi was altered to make him appear to have an extremely long, hooked-shaped nose and the numbers "666" on his forehead. TVS also altered photographs of its own members and those of other white supremacists by incorporating computer-generated artwork (e.g., swastikas, Nazi symbols).

However, many of the CGIs posted on TVS's site were nothing but artwork. For example, computer-generated artwork of Hitler and swastikas appeared with almost the same frequency as the rectangular-shaped banner previously mentioned. In addition to swastikas, the group posted computer-generated artwork of Confederate flags and a host of white supremacist symbols. TVS also posted computer-generated maps of the areas where the group intended to distribute its racist literature. Other CGIs on TVS's site included objects (e.g., spray can), corporate logos, politically oriented cartoons, and artwork depicting whites, Klansmen, and various persons that the group hated (e.g., blacks, Arabs, Mexicans, Mexican-Americans, and Jews).

Although adept in creating CGIs, TVS primarily posted photographic images on its site. Consistent with its interest in self-promotion, TVS posted several "freeze frames" of images that appeared in the video footage that television news stations aired when reporting on the group's offline activities. However, many of the photos on TVS's site were of persons that the group hated, such as President George W. Bush, Jews, blacks, Hispanics, Asians, Africans, homosexuals, members of the press, and whites that the group regarded as traitors to the white race. These photos, which often portrayed the hated individuals in a less than favorable light, sometimes had derogatory terms (e.g., racial slurs) and hateful comments (e.g., "Faggot Freak Show!") incorporated into them.

TVS also posted a variety of photos of persons that the group respected and admired, like Hitler, Nazi soldiers, white supremacists, Klansmen, and its own members. Persons in these photos were shown engaging in activities that included holding guns, playing music, giving Nazi salutes, and making what some consider "obscene" gestures (e.g., raised middle fingers). In some cases, white children as young as toddlers were pictured along with swastikas and other icons that have been used as symbols of hate (e.g., Confederate flags). The group occasionally incorporated computer-generated text (e.g., names, revolutionary rhetoric) into the photos of those persons that it respected and admired.

Allied with NSM at the time of this study, TVS posted photos of the neo-Nazi group's members, compound, and public demonstrations. The NSM demonstration photos posted on TVS's site showed the neo-Nazis, marching down city streets and displaying their symbols of hatred (e.g., swastikas), as well as individuals that were on hand to support or protest the group. TVS also posted photos that were taken at KKK events, such as cross lighting ceremonies and festivals featuring bands playing white supremacist music. The KKK event photos on TVS's site consisted of images of persons attending the events (e.g., Klansmen, band members), the activities taking place (e.g., cross lighting, performances), and the facilities being used, where neo-Nazi and Klan symbols (e.g., swastikas) and artwork (e.g., murals depicting robed Klansmen) could sometimes be seen.

TVS's site also contained photographic images of animals (e.g., dogs), objects (e.g., buildings, gas chambers, weapons), tattooed body parts (e.g., with swastikas), Timothy McVeigh, President George W. Bush, blacks hung from trees, and a dead and dismembered body. Some of the animal photos (e.g., apes, pigs) posted on TVS's site were intended to represent persons that the group hated, such as blacks and police. In some cases, TVS posted a professional news organization's Web page on its site because it included a story about the group's offline activities. These postings contained all of the images originally appearing on the news organization's Web page, such as CGIs (e.g., ads) and various photos, some of which showed flyers that TVS had distributed.

TVS's site included some basic computer-generated animation, namely an image of Hitler moving from side to side. In addition, some racist and anti-Semitic "Skinhead Video Games" were listed as being posted on the group's site, although they could not be accessed at the time content analysis was conducted. Among the video games that were listed on TVS's site were ones that allowed persons to shoot blacks, to fire upon Mexicans crossing the U.S. border, and play a suicide bomber killing Jews in Israel.

TVS also posted a variety of motion video clips, including the notorious—if perhaps unauthentic—footage of Jewish-American Nick Berg being beheaded in Iraq. The group posted this video, which is widely available on the Internet, to celebrate Berg's murder. TVS also posted original video segments that included footage of persons (e.g., Jews, blacks, Mexicans) that the group hated and on-screen comments denigrating those individuals. However, reflecting the group's obsession with self-promotion, the majority of the video clips on TVS's site were stories about the group and its offline activities that aired on professional television news stations. Audio (e.g., music, human voice) was present on all the video footage posted on TVS's site.

## CHRISTIAN IDENTITY

### Orange County Assembly of Christ (OCAC)

While OCAC's Web site did not contain much rhetoric overall, the site did have a fairly diverse array of nontextual content. The group's Web site included a variety of still images, both on its standard Web pages and on printer-friendly PDF-formatted documents that were posted on the site, and even had some basic computer-generated animation.

Like most of the other sampled hate sites, the nontextual rhetoric on OCAC's Web site consisted mainly of photographic images. This is largely because OCAC provided cover shots of the various books that the group had for sale. The covers of these books included artwork, numbers, inanimate objects, persons' names, cartoon-like figures, and a variety of words (e.g., scripture) and icons (e.g., Ten Commandments), both of which were often religiously oriented. The only other photographs on OCAC's site were of the sky and of a white family.

Although the preponderance of the still images on OCAC's Web site were photos, the group also made extensive use of CGIs. The most common CGI on OCAC's site was the group's logo, which was a Confederate flag with images of Jesus Christ and three animals (e.g., eagle, lion, ox) superimposed on it. The majority of the other CGIs on the group's Web site were of a religious nature, such as a Bible, church, and crosses. Other CGIs that appeared on OCAC's site consisted of a dove, buzz saw, artwork, microphones, Confederate flags, other groups' logos, cartoon-like images of inanimate objects, and credit card trademarks used to advertise the fact that individuals could donate to the group. Several of the CGIs on OCAC's site served as hyperlinks to external sites where persons could buy Christian-oriented books or make a cash donation to the group.

The basic computer-generated animation on OCAC's site depicted a black-and-white caricature of a man, who was identified as a Jew, making several different facial gestures and rubbing his hands together as if to suggest that he was money hungry. After the animated sequence ended, the man's image left the screen and was replaced with several terms of oppression (e.g., banned, persecuted, deported), the names of various countries which the group claimed had expelled Jews, and the eras during which these expulsions were supposed to have occurred. All of this was apparently meant to suggest that anti-Semitism is an accepted practice.

## Black Separatist

### Nation of Islam (NOI)

NOI's Web site was very businesslike, as evidenced by the site's design, which included a universal page border that functioned somewhat like a company letterhead. While the content appearing in the middle of the screen changed as one browsed through the site, the borders of the page remained constant. The universal page border consisted of text, links, and still images. What is more, the only nontextual content on NOI's site was still images, usually very basic ones.

The majority of the still images on NOI's site were photographs. Three photos (i.e., exterior of a mosque, head shot of a person, a group of people) were included in the sites' universal page border.

Additional photographs posted on NOI's site consisted of images of natural scenery (i.e., fields, outer space view of the Earth), the interiors and exteriors of buildings, individuals and groups, and important persons in the organization's history (e.g., Elijah Muhammad).

NOI's site also had a handful of CGIs. The universal page border contained some computer-generated artwork and the group's logo, a white star and half moon on a red square background surrounded by a ring bearing two stars, NOI's name, and the words "IN THE NAME OF ALLAH." There were also two natural-computer image hybrids. One of the natural-computer image hybrids was a photograph of Louis Farrakhan incorporated into a graphic that included words and a computer-generated image of a satellite. The other natural-computer image hybrid consisted of photographs of Farrakhan and Jesse Jackson superimposed over a graphic which contained graphic-style lettering and some natural imagery. Other CGIs appearing on NOI's site consisted of flags (i.e., U.S., U.K., Canada) and a banner for *FinalCall.com News*, the group's online newspaper.

**New Black Panther Party (NBPP)**

Unlike NOI's Web site, NBPP's site contained a multitude of nontextual content. Indeed, NBPP's site was among the most sophisticated examined in this study. The group's site included photographic images, CGIs, audio, and motion video.

Like the other sampled sites, NBPP's Web site contained a variety of photographic images. The group's site was composed of photos of animals, objects, structures (e.g., Sphinx), groups of persons (e.g., slaves), portraits of individuals, antiblack literature, proslavery documents, album and magazine covers, and "freeze frames" taken from motion video (e.g., C-SPAN footage of NBPP speeches). However, most of the photos on NBPP's site were of individuals. Consistent with the political theme of its site, NBPP posted photos of blacks that it believed were political prisoners. The group also posted photos of its current and former leaders, persons regarded as leaders in the black community generally (e.g., Al Sharpton, Jesse Jackson), and other blacks that the group respects and admires, such as Rosa

Parks and Louis Farrakhan. At the same time, NBPP's site included photographs of persons that the group hates, including specific whites (e.g., Bill O'Reilly) and Jews (e.g., Ariel Sharon), as well as blacks that the organization regarded as black race traitors, like Oprah Winfrey and Colin Powell.

Several of the photographs on NBPP's site were of the group's offline activities, such as its rallies and public protests (e.g., Million Youth March). Some of these photos showed NBPP members dressed in military-style fatigues and armed with shotguns, both Black Panther Party trademarks dating back to the 1960s. NBPP supplemented its textual rhetoric about police brutality by posting several photographs of blacks who were the victims of such brutality, including the brutalized individuals and their families. For example, the group's site included photos of Amadou Diallo, the Haitian immigrant who was shot 41 times by New York City police officers. NBPP also posted photos of police officers, both black and white, who had been accused of brutalizing blacks. In addition, the group posted photographic images of individuals protesting the police, police-protester clashes, and altercations between police officers and black citizens. Some of these images appeared to have been "freeze framed" from television news video. NBPP incorporated graphic-style lettering into several of the photos that were posted on its site.

NBPP's site also contained a multitude of CGIs, nearly equal to the number of photographic images that the group posted online. Many of the CGIs on NBPP's site were natural-computer image hybrids. For example, there was a photo of John Ashcroft (the then-current U.S. Attorney General) that had been modified to make him appear to have a Hitler-style mustache. Additional natural-computer image hybrids on NBPP's site included photographic images that had been combined with computer-generated artwork (e.g., George W. Bush's face superimposed over a Nazi flag).

However, the most prevalent CGI on NBPP's site was the group's logo, which consists of an image of a black panther, superimposed over artwork depicting Africa, surrounded by two rings, one of which bears the group's name, two stars, and the words "FREEDOM

OR DEATH." The computer-generated logo appeared both by itself and as part of other CGIs on the site, such as a rectangular-shaped banner that appeared at the top of many of the group's hate pages. Like the group's logo, many of the CGIs posted on NBPP's site were a combination of computer-generated artwork and text. Additional CGIs posted by NBPP included artwork of "Uncle Sam," various versions of a black panther, and a cartoon-like spider that appeared in the "Panther Cubs" children's section of the group's site.

There were also a variety of audio files posted on NBPP's site. For one, there was an audio transcript of an episode of the *Donahue* television show in which the featured guest was Dr. Khalid Abdul Muhammad, NBPP's former leader. On the transcript, Muhammad and some members of the in-studio audience could be heard expressing, among other viewpoints, problack, antiwhite, anti-Jew, and anti-American sentiments. NBPP's site also contained a variety of profanity-laden gangster rap songs, most of which had antiwhite (e.g., "Kill Whitey") and anti-American (e.g., "Amerikkka's Most Hated") themes.

NBPP supplemented its textual rhetoric regarding police brutality by posting CBSNews.com videos of altercations between white police officers and black individuals. One of these video clips also included an interview, a CBS promotional advertisement, and a Wal-Mart commercial, all three of which contained audio (e.g., music, human voice).

## NEO-CONFEDERATE

### League of the South (LOS)

LOS's hate site also contained a variety of nontextual rhetoric, which, like NBPP's Web site, included photographic images, CGIs, and audio. However, while LOS's site did not contain any motion video, it did have some basic computer-generated animation.

The photographic images on LOS's site largely consisted of photos of individuals, including columnists, political writers, Abraham Lincoln, and the group's own members, in particular its president, Dr. J. Michael Hill. The group also posted photos of persons attending its national conferences and public demonstrations. The conference photos on LOS's site showed individuals conversing, delivering speeches, playing musical instruments, assembled together in large groups, browsing through merchandise being sold at the events, and performing what appeared to be a skit about the Confederate South. The demonstration photos showed individuals delivering speeches, playing musical instruments, marching down public streets, gathering at county courthouses, waving Confederate and Southern states' (e.g., Georgia) flags, and holding banners inscribed with Southern symbols and religious rhetoric (e.g., "no king but Jesus"). Many of the individuals that appeared in the demonstration photos on LOS's Web site were wearing Confederate-style military uniforms. The group indicated that some of the photos of persons on its site originally appeared in professional newspapers (e.g., *The Sun Herald*). LOS incorporated computer-generated text into some of the photos of individuals it posted online.

Many of the photographic images on LOS's site were of the merchandise that the group, or one of its chapters, had for sale. The group posted photos of a variety of different kinds of LOS-brand merchandise, such as postcards, books, music CDs, LOS name badges and membership certificates, and audio tapes of lectures that took place during the group's national conferences. In addition, LOS posted photographic images of non-LOS merchandise that it encouraged others to purchase, such as books and portraits of Confederate leaders (e.g., Robert E. Lee, "Stonewall" Jackson). LOS's site also included photographic images of columnists, buildings, inanimate objects (e.g., coin, stamp), the group's own written documents, Confederate soldiers, and portraits of historical Southern figures (e.g., Jefferson Davis) and blacks living in the Confederate South. Some of the photographic images on LOS's site served as hyperlinks to external Web sites, usually pro-South ones.

Like AWKKKK's Web site, LOS's site contained more CGIs than photographic images. The most common CGI on LOS's site was an icon, which incorporated in its design a man, Confederate flag, and graphic-style lettering that formed the group's Web site address (i.e., www.dixienet.org). This CGI appeared at the bottom of virtually every one of the group's hate pages. There were several other CGIs on LOS's site in which the Confederate flag was either incorporated (e.g., in a dog's mouth, draped over an image of Florida) or appeared by itself. LOS also posted CGIs of a variety of other flags, including the Confederate States of America (CSA) flag, which was adopted by the Confederate States Congress in 1865, and the flags of Canadian provinces (i.e., Quebec), several Southern States, and other LOS chapters.

As a reflection of its pro-South sentiment, LOS promoted Southern businesses and other pro-South groups by posting CGIs of their logos. LOS also posted CGIs of products being sold by various businesses, both Southern and non-Southern. Other CGIs appearing on LOS's site included maps (e.g., U.S., of the South), the group's own logo, Abraham Lincoln, Southern plantations, natural-computer image hybrids, and a variety of other computer-generated artwork (e.g., scroll, Confederate soldiers). Some of the CGIs on LOS's site served as hyperlinks to external sites, such as those operated by pro-South groups and Southern businesses.

The audio on LOS's site consisted of an archive of samples of lyric-free, Southern-style music (e.g., bluegrass). The basic computer-generated animation on the group's site depicted a waving CSA flag.

## Other Hate Groups

### Council of Conservative Citizens (CCC)

Just as CCC is conservatively inclined, so too was the group's Web site conservatively designed. The only nontextual content posted on the group's site was still images, mainly unsophisticated looking ones.

Like the sites operated by AWKKKK and LOS, CCC's hate site contained more CGIs than photographic images. Yet, the CGIs appearing on the group's site still tended to be very basic. For example, the most recurrent image on the site was CCC's logo, which primarily consists of the group's name written in graphic-style lettering. The group demonstrated its affection for symbols of Southern heritage by posting various CGIs of Confederate flags. Additional CGIs on CCC's site consisted of a mailbox, typewriter, globe, gear box, natural-computer image hybrids, a caricature of a man looking through binoculars, and an old-fashioned black-and-white drawing of signals emanating from an antenna sitting atop the earth's surface. Several of the CGIs on the group's site had words incorporated into them or served as hyperlinks to external Web sites, such as a pro-Confederate flag site.

CCC's site also contained several photographic images. Many of the photographs on the site were of people attending CCC's meetings and public demonstrations (e.g., anti-immigration, proflag). The meeting photos showed persons speaking, sitting, standing, dancing, playing musical instruments, and walking toward a building. The public demonstration photographs showed persons posing for a picture, raising Mississippi's state flag, and holding signs expressing proflag and anti-immigrant sentiments. CCC also expressed its anti-immigrant sentiment visually by posting photos of what it claimed were white South Africans who had been murdered by their black counterparts. The organization suggested that these images provided support for its position that the immigration of nonwhites into the United States poses a threat to white Americans. Authentic or not, the photographs, which were most likely downloaded from an external site, were still gruesome images of dead men, women, and children that appeared to have either been shot, burnt, strangled, drowned, or dismembered. CCC's site also included photographic images of a statue, book covers, the city of St. Louis, a *Washington Times* article, the organization's newspaper, some of the group's members, and an anti-immigration placard situated in what appeared to be a city park. Some of the photographic images on the group's site served as hyperlinks to external Internet sites, like a Web site that sold Southern-themed merchandise.

**Jewish Defense League (JDL)**

JDL may be a radical Jewish organization, but the group's Web site was anything but radical in terms of its design. What nontextual rhetoric was on the group's site consisted mainly of photographic images.

The majority of photographic images on JDL's site were of individuals. The group posted images of various individuals, such as its members and other persons, mainly Jewish, that the group indicated that it respected, admired, or had committed itself to defending the rights of, such as a slain black girl. However, JDL is a hate group and as such, the organization also posted photos of persons that were the objects of its hatred, including members of neo-Nazi groups and other individuals that the group accused of being anti-Semitic (e.g., director Michael Moore, comedian Jay Mohr). In some cases, JDL posted photos of the buildings and facilities used by neo-Nazi groups (e.g., Aryan Nations) or the homes and places of employment of those individuals that the group regarded as "Jew haters."

Several of the photos on JDL's site were of the group's public protests. JDL's public protest photos showed law enforcement officials, media representatives, the protested hate groups delivering their speeches, and protesters engaged in activities ranging from merely congregating together to burning Nazi flags to displaying antihate (e.g., swastika encircled by a prohibited symbol) and pro-Jewish (e.g., Star of David) symbols. Demonstrating its interest in raising awareness about acts of Jew hatred, the group also posted photographs of literature (e.g., comic strips) and artwork (e.g., sports team logos, murals on city buildings) that it believed were anti-Semitic in nature. Other photographic images on JDL's site included art, signs, posters, buildings, JDL merchandise (e.g., T-shirts), and letters mailed to the organization.

JDL's site did, however, contain some other forms of nontextual content, including some CGIs. The most common CGI on JDL's site was the organization's logo, a clenched fist inside the Star of David, which appeared at the top and bottom of virtually every one of the group's hate pages. CGIs also appeared in the various articles that JDL posted on its site. For example, an article about an event being sponsored by the neo-Nazi group Blood and Honour contained the group's logo (i.e., the words "Blood" and "Honour" surrounded by artwork). As part of its efforts to promote its public protests, JDL posted several computer-generated road maps that directed persons to those areas where the group was planning to hold a demonstration. JDL's site also contained CGIs of graphs, phones, faxes, mailboxes, notepads, envelopes, corporate logos (e.g., Google, Estée Lauder, Yahoo!), and a variety of other artwork. There was even a natural-computer image hybrid on JDL's site in which a photo of Pat Buchanan had been altered to make him appear to have a Hitler-style mustache. Buchanan's image, along with the statement "Let's Finish Hitler's Job," was superimposed over computer-generated artwork depicting the White House. Several of the CGIs on JDL's site had graphic-style lettering incorporated into them.

There was one clip of motion video posted on JDL's site as well. The video was from an episode of *The Jerry Springer Show* in which some JDL members appeared along with Klansmen as in-studio guests. The video clip also contained the audio that aired during the telecast.

## Summary

As Breckheimer (2002) writes, "because of the capabilities associated with the Internet, hate groups are no longer limited to text messages" (p. 1495). Realizing this, U.S.-based hate groups have taken full advantage of the nontextual capabilities that the Internet has to offer. Indeed, this chapter revealed that each of the 10 U.S.-based hate sites examined in this study contained several different forms of nontextual content.

Not surprisingly, all 10 of the sampled hate sites contained still images, as do most Web sites. What is more, like most Web sites, still images were the main form of nontextual content on each of the sites. Along with still photographic images (imagery that existed apart from a computer that can simply be reproduced in an Internet-ready digital format), each of the sampled sites contained a variety of stationary images that were completely computer-generated, further evidence that hate groups have become more computer-savvy. Indeed, three of the hate sites (AWKKKK, CCC, LOS) included more computer-generated images than photographic images. What is more, many of the computer-generated images appearing on the sampled sites were rather sophisticated in terms of their design.

The next most common forms of nontextual content found on the sampled hate sites were computer-generated animation and audio, both of which were present on six Web sites. Like many of the computer-generated images appearing on the sites, some of the computer-generated animation was rather sophisticated, although most tended to be very basic. Interestingly, the only Web sites containing computer-generated animation were those maintained by hate groups whose members, it would seem, are (or were) all white (AWKKKK, LOS, MWKKKK, NSM, OCAC, TVS). However, audio was found on Web sites operated by a variety of hate groups (AWKKKK, JDL, LOS, NBPP, NSM, TVS), both white and nonwhite. In addition to music (the main type of audio-based content present on the sampled sites), there were also sound effects (e.g., thunder, dinosaur roaring) and audio of persons talking.

Three of the sampled hate sites (JDL, NBPP, TVS) even contained motion video, the majority of which was originally produced by professional media organizations (e.g., television news stations, television talk shows). Most of this motion video also included audio (e.g., music, human voice).

# Chapter 10

# Findings

This chapter examines whether the content found on the 10 sampled sites was constitutionally proscribable under the First Amendment tests determined to be appropriate for assessing the constitutionality of online hate speech: "*Miller*," "true threats," "clear and present danger," and "bad tendency." A categorical balancing approach was used to determine whether content on the sampled sites could be placed into an existing category of unprotected speech as defined in these four First Amendment standards developed by the Supreme Court.

## The "*Miller* Test" for Obscenity

There was nothing on the 10 sampled hate sites that would have been considered obscene under the *Miller* three-part test for obscenity. Not only was there no nudity on any of the sites, there was also no content, textual or nontextual, that depicted or described "sexual conduct," let alone in a "patently offensive way." Thus, none of the content on the sampled sites satisfied the *Miller* test's second prong, which, along with the other two parts of the test (i.e., "appeals to the prurient interest," "lacks serious literary, artistic, political, or scientific value") must be met in order for materials to be deemed legally obscene.

While many of the hate sites contained profanities that are also used to refer to sexual conduct, the Supreme Court established in *Cohen v. California* (1971) that these sorts of expletives cannot render nonsexual content legally obscene. Add to this that each Web site contained an extensive amount of political or religious content (and oftentimes both), and it becomes difficult to see how any of the sampled sites, "taken as a whole," would have ever been regarded as legally obscene by a court of law.

## THE *BLACK* COURT'S "TRUE THREATS" TEST

There was also no content on any of the sampled hate sites that would have been regarded as constituting a "true threat" under the guidelines set forth by the Supreme Court in *Virginia v. Black* (2003). Although the hate groups in this study expressed hatred for various persons, nowhere on their Web sites did they direct any threats of "unlawful violence" toward any "particular" "individual," or "group" small enough that each person belonging to it could be affected as an individual.

First and foremost, there was no explicitly threatening language directed toward any particular individual or group of individuals on any of the sampled sites. The closest phrase to explicitly threatening language on the sites was a statement posted on JDL's site in which the Jewish organization indicated that it would subject certain persons to its "wrath" if they did not apologize for being anti-Semitic. However, viewed in context, which the Supreme Court has repeatedly held must be considered along with content when assessing whether speech constitutes a "true threat," even this statement cannot reasonably be construed as having been a threat to commit unlawful violence. This is due to the simple fact that JDL never expressed an intention to perpetrate violence anywhere on its site. Viewed in the context of the group's online rhetoric and offline activities, the "wrath" to which JDL referred would more reasonably be interpreted as the way in which the organization angrily expressed its opposition through boycotts and public protests to persons it regarded as being anti-Semitic.

Moreover, there was no content, textual or nontextual, on any of the sampled sites that would have been regarded as being posted "for the purposes of" merely implying to any particular individual that she or he had been targeted for unlawful violence. It is true that some of the sites listed individuals' names, and sometimes even their photos and contact information, which, like an explicit threat, could have placed the identified persons "in fear of" being victimized by unlawful violence. However, there was no basis for concluding that any of these postings constituted a "true threat." To begin with, one would have had no way of knowing if an individual's name, photo, or contact information was listed for the purposes of placing the identified person in fear that he or she was the target of unlawful violence. Of course, one can assume that any group in this study would have known that an individual would likely interpret his or her being identified on its site as a "true threat" if persons were routinely subjected to unlawful violence after being identified in its materials. Yet, there was no evidence to suggest any pattern of unlawful violence being perpetrated against persons who had been identified in any materials produced by any of the hate groups in this study, or indeed any U.S.-based hate groups which shared their hateful ideologies. In *PPCW*, the Ninth Circuit Court of Appeals held that the listing of abortion providers' names on an antiabortion Web site qualified as a "true threat" primarily because there was a previous pattern of violence against abortion providers who had been identified in antiabortion materials. Without knowledge of intent or of any pattern of violence against persons identified in materials produced by the studied groups or other like-minded groups, it was concluded that the identification of persons on the sampled sites alone did not constitute a "true threat" as defined by the *Black* Court. The lack of any explicitly threatening language on any of the sites certainly supported this conclusion.

At most, some of the sampled sites could be said to have contained content that implicitly threatened groups so large that there was simply no way that all persons belonging to them could have been affected as individuals. However, the *Black* Court's "true threats" test cannot be applied to threats of this sort. On NSM's neo-Nazi Web site it was stated that all Jews, nonwhites, and child molesters should be killed. A gangster rap song (i.e., "Kill Whitey") posted on the site maintained by the militant black separatist group NBPP contained lyrics suggesting that blacks should murder whites, who are accused in the song of being antiblack. TVS's racist skinhead site included images depicting, and stories describing, violence (e.g., hanging, shooting, dragging behind a car) being perpetrated against Jews, foreigners (e.g., Mexicans), and nonwhites (e.g., blacks). For example, there was one story posted on the group's site about whites shooting Mexicans, both adults and children, as they tried to cross the U.S. border illegally. However, this story was labeled as "fiction," as were most of the other stories discussing violence on TVS's site. Without evidence to the contrary, there was no reason to believe that all the stories of violence posted on TVS's site were anything more than savage and tasteless pieces of fiction, to borrow a phrase from the U.S. District Court in *United States v. Baker* (1995). TVS's story about whites shooting Mexicans, and the other hateful rhetoric just mentioned, appeared to be some hatemongers' way of directing threats of violence at individuals belonging to very large groups without running any real risk that a court would ever find that the communications were constitutionally proscribable "true threats."

## "Clear and Present Danger" and "Bad Tendency" Tests

The research also did not uncover any content on the sampled sites that satisfied either the "clear and present danger" or "bad tendency" tests, both of which were designed for evaluating speech considered likely to pose a threat to government. While some sites included commentary critical of governmental laws, programs, policies, and officials, none contained any explicit threats directed toward any U.S.-based governmental body or representative. What is more, none of the sampled sites contained any discourse advocating that persons should defy—let alone attempt to overthrow by force or violence—any form of government in America.

Indeed, the rhetoric that some of the groups posted on their sites even tended to suggest that they were progovernment. In promoting a rally that it planned to hold at a county courthouse to generate support for additional child molestation laws in Mississippi, MWKKKK informed potential attendees that government officials had indicated that anyone attending the demonstration must stay off the courthouse lawn. What is more, the group expressed its support for American military troops and any other decisions or actions that would help the United States remain a powerful nation. The content of the Web site operated by the white conservative group, CCC, indicated that the organization is also progovernment. For example, even as it criticized President George W. Bush for supporting a plan that would grant amnesty to millions of illegal immigrants residing in the United States, the group provided links to pro-Bush sites.

Although not progovernment, several of the other groups in this study, including those that were decidedly antigovernment, indicated (some perhaps insincerely) that they had no intention of defying government through any sort of lawless action. One of the black separatist groups, NOI, expressed its belief that persons should respect the laws of the land and never become aggressive or make war. In addition to claiming that it adhered both to a policy of lawfulness and nonviolence, AWKKKK advised its supporters not to violate the law at various points on its site. As an example, in advertising a protest rally that it eventually held at a county courthouse in Georgia, the group posted some of the state's public demonstration ordinances along with a request that any individuals wishing to violate any of the statutes refrain from attending the event.

The neo-Nazi organization, NSM, and the racist skinhead group, TVS, also advised their supporters against operating outside the law. Mindful that government officials would be provided with a reason to investigate, and perhaps prosecute, the group if it were linked to illegal activity, NSM told its supporters that they should only attempt to further the group's cause through legal means. As an example, NSM encouraged its supporters to distribute the group's Nazi literature. Along with informing persons about how they could legally support its objectives, TVS reminded its followers that incarcerated white supremacists could not help the group in its fight to preserve the white race. Even the neo-Confederate organization, LOS, which repeatedly expressed the belief that the Southern states should consider seceding from the United States, claimed it was committed to trying to effect change in local, state, and federal government through lawful political means. For example, the organization encouraged individuals to vote for certain candidates and legislation that it endorsed.

Similar to what was previously stated with regard to the hateful person-centered rhetoric on the sampled sites, at most it could be argued that some of the Web sites in this study contained content that, when viewed in isolation, implicitly encouraged others to perpetrate illegal acts that could threaten government. NSM's neo-Nazi site included some content of a militaristic nature, such as U.S. Army field manuals and rhetoric on the importance of military leadership. The racist skinhead group, TVS, had two hate pages on which it endorsed the practice of "lone wolfing" and "leaderless resistance," terms which (as previously noted) are commonly used to denote the concept of persons independently carrying out a group's call to violence. Depending on the group, a call to violence could include acts which could potentially threaten government. However, just as the First Amendment requires that contextual factors be taken into consideration when deciding whether speech constitutes a "true threat" to individuals, so too does it require that context be considered when assessing whether speech represents an actual threat to government.

When viewed in context, even the military-oriented content on NSM's Web site and the rhetoric about "lone wolfing" and "leaderless resistance" on TVS's site did not fall into the "threatening-to-government" category of unprotected speech as generally defined by the "clear and present danger" and "bad tendency" tests. Nowhere did NSM indicate that it posted the military-oriented content on its site to encourage attacks against any existing U.S.-based governments. Rather, the group indicated that it posted the material on its site in order to provide white Americans with information on how to defend America, themselves, and their interests in the event that the U.S. government falls. Moreover, even if the terms "lone wolfing" and "leaderless resistance" are usually used to denote the concept of persons independently carrying out a group's call to commit violent acts, the terms were used to refer to something else on TVS's site. TVS used the terms "lone wolfing" and "leaderless resistance" to denote the concept of individuals endeavoring to advance the white supremacist cause on their own, or as part of a small group, rather than as members of a large, structured white supremacist organization. TVS advised its fellow white supremacists that this sort of "lone wolfing" and "leaderless resistance" was the best way for them to avoid being subjected to surveillance by government, which the group believed was inclined to monitor anyone suspected of holding white supremacist views. Nowhere in its discussion about "lone wolfing" and "leaderless resistance" did TVS ever mention violence or any other kind of illegal activity. Thus, when viewed in context, neither NSM's military-oriented content nor TVS's rhetoric about "lone wolfing" and "leaderless resistance" appear to have been posted for the purposes of implicitly encouraging others to commit illegal acts, let alone any that would threaten government.

## Summary

While some countries have banned Internet hate sites, if the findings of this study are any indicator, there is no basis in American free speech jurisprudence for the United States to take similar measures. This study found that the publicly accessible content of the 10 sampled Web sites was not constitutionally proscribable under any of the Supreme-Court-devised First Amendment tests that were determined to be the most appropriate standards for assessing the constitutionality of hate sites: "*Miller*," "true threats," "clear and present danger," and "bad tendency."

To begin with, this study did not find any textual or nontextual content on the sampled sites that depicted or described sexual conduct, the threshold requirement that speech must meet in order to be considered legally obscene under the *Miller* test for obscenity. Even though several of the sampled sites contained profanities that are slang for sexual conduct, the Supreme Court has already established that these kinds of expletives cannot render otherwise nonsexual content legally obscene.

What is more, this study did not find any content on the sampled hate sites that would have been regarded as constituting a "true threat" within the meaning of the *Black* Court's test. For one, there was no explicitly threatening language directed at any particular persons on any of the sites. Additionally, there was no content, textual or nontextual, that would have been regarded as being posted for the purposes of implying to any particular individual that he or she had been targeted for violence. At most it could be argued that some sites contained content that implicitly threatened groups so large that it was impossible that all persons belonging to them could have been individually affected. However, these types of threats are outside the scope of the *Black* Court's "true threat" test.

Lastly, this study did not find any content on the sampled Web sites that satisfied either the "clear and present danger" or "bad tendency" tests, both of which were designed to provide guidelines for assessing the constitutionality of speech thought likely to pose a threat to government. First, none of the sites contained explicit threats directed toward any governmental entity based in the United States. Moreover, each Web site was devoid of discourse advocating that persons attempt to forcefully or violently overthrow, or defy through lawless action, any form of government in America. These were precisely the forms of advocacy that the "clear and present danger" and "bad tendency" tests were manifestly intended to apply, as shown by the context in which they were formulated. Two sites

(NSM, TVS) contained content that, if viewed in isolation, might have been regarded as encouraging persons to commit illegal acts that would threaten government. However, when these materials were viewed in context, which the First Amendment requires, they did not appear to have been posted for that purpose.

# Chapter 11

# Summary and Conclusions

This study was organized around the central question: Are online hate sites deserving of First Amendment protection? The findings of this study strongly suggest that if the content appearing on the publicly accessible areas of the 10 sampled Web sites is indicative of that generally found on other U.S.-based hate sites, then the answer to that question is a resounding "yes."

This study's conclusion regarding the constitutionality of hate sites is consistent with that drawn by Leets (2001), who determined that the Web site rhetoric of white supremacists was protected expression. However, this study's conclusion is even more significant because, unlike Leets' content analysis, this study was not restricted to hate sites posted by white supremacist groups. Indeed, the sample of hate sites examined in this study consisted of Web sites posted by every type of U.S.-based hate group identified by the Southern Poverty Law Center, a leading monitor of the activities of hatemongers. That sample included Web sites posted by the following types of hate groups: "Ku Klux Klan" (2), "neo-Nazi" (1), "racist skinhead" (1), "neo-Confederate" (1), "Christian Identity" (1), "black separatist" (2), and "other" (2). The two Web sites from the "other" hate group category included a site maintained by a white conservative organization and one of the few U.S.-based hate sites operated by a Jewish group. Thus, the sample used in this study is more representative of the population of Internet hate sites than are those samples limited to Web sites maintained by groups with white supremacist ideologies. As a result, the determination that the sampled Web sites warranted First Amendment protection is more likely to be an accurate assessment of the constitutionality of the population of online hate sites in general.

Given that the evidence from this study and similar studies suggests that online hate sites are deserving of First Amendment protection, alternatives to governmental speech regulation need to be considered for regulating hateful communications posted on the publicly accessible areas of Web sites. User-based Internet filtering technologies are the most appropriate nonlegal remedy for regulating hate sites, and indeed all online content that, while perhaps harmful to minors, is constitutionally protected for adults. Although most legal commentaries regarding filters center around their ability to screen out sexually oriented Internet content, several First Amendment scholars have also discussed the issue of whether filtering technologies might serve as an alternative to government regulating hate speech posted online.

The stage has already been set to allow user-based filtering technologies to function as an alternative to governmental regulation of Internet hate sites. Many commercially available filters have been designed to allow users to block out Web pages that contain "hate speech," as defined by the respective software companies. As an example, CyberPatrol's filtering technology, "CyberLIST," blocks access to Web sites that, in the company's judgment, attack "specified populations or institutions based on associations such as religion, race, nationality, gender, age, disability, or sexual orientation" (CyberPatrol, n.d.). Additionally, several organizations offer free hate-filtering software on their Web sites (Timofeeva, 2003). Like filtering technologies used to block out sexually explicit Internet content, "hate filters" allow persons to restrict their access to certain

Web pages on the basis of particular words (e.g., kike), addresses (e.g., www.kkk.com), or other information used to identify Web page content. "Hate filters" are also similar to other online filtering technologies in that they allow users to adjust the software's screening criteria to reflect their own filtering preferences, which may be more or less restrictive than the blocking criteria originally established by the company that designed the software.

Some critics of filters point out that the technology is not entirely effective, and this is certainly a fair assessment. On one hand, filters occasionally have problems with "underblocking," letting in Internet content that they were supposed to filter out. On the other hand, filters sometimes have problems with "overblocking," filtering out Internet content that they were supposed to let in. User-based hate filters have these same problems.

Although hate filters are not perfect, and likely never will be, they remain preferable to any regulation of hateful online material that might be imposed by government. As noted at the outset of this book, many scholars have observed that the Framers' First Amendment conception of free speech was based on the belief that all viewpoints should compete for adherents in a "marketplace of ideas" free from government interference. By allowing individuals, rather than government, to control what content they or minors in their custody have access to on the Internet (an electronic marketplace of ideas frequented by millions of persons), user-based hate-filtering technologies would be able to foster the free-market conception of the First Amendment.

It is true that hate filters will sometimes let in content from the online marketplace that certain individuals would prefer not be confronted with, or occasionally block out Internet materials that some persons would like to access. However, these same sorts of problems would also result if the U.S. government attempted to regulate hate sites. This is due to the fact that like much of the terminology used in the area of free speech regulation (e.g., "indecent," "obscene," "patently offensive"), "hate" is a term that defies universal definition when applied to specific content. What one individual regards as an obvious form of "hate speech," others may simply view as critical discourse or an expression of self-satisfaction in being affiliated with a certain group. For example, some would view a Southern-based white conservative organization's display of the Confederate flag on its Web site as the means by which the group communicates a message of racial intolerance to nonwhites, while others would consider it a symbolic expression of Southern pride. Thus, if the U.S. government attempted to regulate hate sites, its regulatory efforts would inevitably be viewed as "underblocking" by some, and "overblocking" by others. After all, any governmental regulation of hate sites would still involve persons making subjective determinations as to what sorts of Web site content actually constituted "hate speech," just like what is involved in setting a hate filter's screening criteria. However, unlike what happens when user-based hate filters "overblock" Web site content at the point of reception, a government-imposed ban against posting certain content on the Internet would remove that material from the online marketplace of ideas altogether.

In short, user-based filters are the most appropriate means of regulating hate sites because they help strike a balance between the First Amendment rights of individuals to post and access hate speech online and the interests that others may have in restricting their own access to this type of discourse. Of course, some people are skeptical of the notion of relying on technology to serve as an effective alternative to governmental regulation of speech. However, persons may have to accept the fact that a user-based filtering regime currently represents the most appropriate regulatory scheme for dealing with hate speech posted online. As Weintraub-Reiter (1998) writes, "The Supreme Court established the need to find means of filtering hate speech when it granted the Internet full constitutional protection" (p. 158).

Nevertheless, should the U.S. government find it necessary to regulate online hate sites, this study makes an important contribution to the ongoing debate about what First Amendment tests it should, and should not, draw upon in devising its regulatory scheme. This study finds that of the Supreme Court's six First Amendment standards for evaluating the constitutionality of hateful speech communicated in non-Internet settings, only half are applicable to hate sites. Specifically, the study finds that the Court's "fighting words," "group defamation," and "imminent lawless action" tests are inapplicable to hate sites, while its "clear and present danger," "bad tendency," and "true threats" tests are applicable. In addition, the study finds that the Court's "*Miller* test" for obscenity could also be applied to hate Web sites, even though the standard was not specifically designed for assessing hateful speech. Although none of the sampled Web sites contained obscene content, there always exists the possibility that U.S.-based hate sites could include material that rises to the level of obscenity. Because of their applicability to hate sites, the Supreme Court's "*Miller*," "clear and present danger," "bad tendency," and "true threats" tests were utilized to assess the constitutionality of the Web sites selected for inclusion in this study.

However, while this study's primary purpose was to assess the constitutionality of online hate sites, the analysis of the content of the sampled Web sites produced information of relevance to scholars, especially Internet and mass media researchers. This is largely due to the fact that relatively few scholars have examined hate site content. For one, this analysis revealed the thematic content of the various Web sites examined in this study. This information sheds light on the ideological underpinnings of the hate groups that maintained the sampled sites while also providing insight about the beliefs that other like-minded hatemongers may hold, and perhaps be discussing online. Several themes, both major and minor, appeared in the rhetorical discourse found on the sampled sites, and, contrary to what the term "hate site" may imply or what some persons might expect from a "hate group," they were not always hateful in nature. In fact, one of the most significant findings of this study was that political and religious speech was a major component of the sampled hate sites.

What is more, some of the themes that emerged were common to several of the sampled sites, which indicates that many hate groups, including those operating from differing ideological perspectives within the hate movement, have very similar beliefs. Of the major themes common to multiple Web sites, white supremacy was a theme of the neo-Nazi, racist skinhead, Christian Identity, white conservative, and both KKK sites, while racial separatism was a theme of the Christian Identity, both black separatist, and both KKK sites. The analysis of the thematic content of the sampled sites also revealed the various rhetorical techniques that the hate groups in this study employed on their Web sites, information which also contributes to rhetorical theory regarding hate group, social movement, apocalyptic, and Internet rhetoric.

Additionally, this study provides a better understanding of hate sites through its identification of the various persons that were the targets of hatred on the 10 sampled sites. The hate groups in this study expressed animosity toward a vast array of persons and groups of persons on their Web sites. However, while particular individuals were the objects of hatred on all but two of the sites, the majority of the hateful person-centered rhetoric on the sampled Web sites was directed at groups of persons, typically very large groups. Jews and homosexuals were the most common targets of hatred, each group being denigrated on a total of seven Web sites, although the anti-Jew rhetoric appearing on the sites was generally more vicious and voluminous. That anti-Jew rhetoric appeared on a majority of the sampled sites was not all that surprising given that Jews have steadily become the primary target of the hateful rhetoric of U.S.-based hate groups, especially Klan and neo-Nazi organizations. However, it was surprising that homosexuals were disparaged on an equal amount of the sampled sites. The high volume of antihomosexual rhetoric on the sampled sites was most likely a negative consequence of the increasing visibility of gays and lesbians and the growing acceptance of the homosexual lifestyle in American society.

The next most frequently targeted group was nonwhites, who were the objects of hatred on half of the sampled Web sites, which is not all that surprising given that most of the hate sites in this study were maintained by whites. What is surprising is that in spite of 9/11, Arabs and Muslims were disparaged on less than half of the sampled sites, as were a variety of other groups. Also surprising was that the hate groups in this study frequently expressed animosity toward persons for reasons other than their sex, race, ethnicity, religion, or sexual orientation, the innate human characteristics generally associated with the term "hate speech." Indeed, some of the sampled hate sites did not include any rhetoric denigrating persons on the basis of these characteristics.

Information regarding the persons to which hatred is leveled against on hate sites is of value to researchers. The information lends further insights into the organizations producing this hateful rhetoric, how certain events and circumstances may have influenced the discourse that they created, and the potential effects that this content could have on those persons who are the targets of these groups' hatred. Moreover, these findings contribute to researchers' understanding of ways in which Internet content can potentially dissuade certain groups from going online, inhibit certain types of discourse in cyberspace, and perhaps even exacerbate problems in the offline world.

Lastly, additional knowledge was produced by establishing the nontextual modes of presentation, visual and audio-based, that the hate groups in this study utilized on their hate sites. Each of the sampled hate sites contained several different forms of nontextual content. Not surprisingly, all of the sampled sites contained still images, as do most Web sites. What is more, like most Web sites, still images were the main form of nontextual content on each of the sites. However, in addition to containing still photographic images (imagery that existed apart from a computer that can simply be reproduced in an Internet-ready digital format), each of the sampled sites contained a variety of completely computer-generated stationary images. The existence of computer-generated images on all the sampled sites provides further evidence that hate groups have become more computer-savvy. The next most common forms of nontextual content found on the sampled hate sites were computer-generated animation and audio, both of which were present on six Web sites. Interestingly enough, while the only Web sites containing computer-generated animation were those maintained by whites, audio-based content (mainly music) was found on sites operated by various hate groups, both white and nonwhite. Three of the sampled hate sites even contained full motion video.

The findings regarding the nontextual modes of presentation used on the sampled hate sites are important. This information helps to reveal the underlying motivations of those persons who posted this mass-mediated content and the possible effects the various presentational styles may have on Internet audiences. The wide variety of nontextual content present on the sampled sites strongly suggests that the hate groups in this study, and other hatemongers that have an online presence, are trying to appeal to a wider array of people, including youth and children. Specifying the nontextual modes by which the content of hate sites is presented is also important because research has shown that the effect that mediated messages have on audiences is partly due to whether they are delivered in the form of audio or visuals.

However, for all the contributions that this study makes, the issue and the content of U.S.-based hate sites demand further examination—especially by legal and communication scholars. All of the evidence suggests that the number of these Web sites, which has increased several hundred-fold since the first reported U.S.-based hate site was posted online in 1995, will continue to proliferate. What is more, the findings of this study indicate that the U.S. government presently has no basis under existing First Amendment law to stop this phenomenon from continuing. Indeed, unpopular speech, like that found on hate sites, is exactly the sort of speech the First Amendment is supposed to protect, since popular speech needs no protection. So, persons offended by hate sites should utilize the available means for avoiding exposure to unwelcome Internet content rather than relying on government to restrict the free speech rights of those who created these controversial Web sites. Because like it or not, freedom of speech is a right that extends to all Americans, even hatemongers.

# References


Abrams v. United States, 250 U.S. 616 (1919).

Ackerman, G. A. (2003). Beyond arson? A threat assessment of the Earth Liberation Front. *Terrorism and Political Violence, 15*(4), 143–170.

ACLU v. Gonzales, 478 F. Supp. 2d 775 (E.D. Pa. 2007).

ACLU v. Johnson, 194 F.3d 1149 (10th Cir. 1999).

ACLU v. Reno, 929 F. Supp. 824 (E.D. Pa. 1996).

American Booksellers Foundation v. Dean, 342 F.3d 96 (2d Cir. 2003).

American Coalition of Life Activists v. Planned Parenthood of the Columbia/Williamette, 539 U.S. 958 (2003).

American White Knights of the Ku Klux Klan (AWKKKK). (n.d.). *New page 1*. Retrieved December 1, 2004, from http://www.awkkkk.org/

Andrews, J. R. (1980). History and theory in the study of the rhetoric of social movements. *Central States Speech Journal, 31*(4), 274–281.

Apple, A. L., & Messner, B. A. (2001). Paranoia and paradox: The apocalyptic rhetoric of Christian Identity. *Western Journal of Communication, 65*(2), 206–227.

Ashcroft v. ACLU, 535 U.S. 564 (2002).

Ashcroft v. ACLU, 124 S. Ct. 2783 (2004).

Barnett, B. A. (1996). *Hate speech, free speech, and the Internet: Toward an access model of regulation.* Unpublished master's thesis, University of Akron, Ohio.

Barron, J. A., & Dienes, C. T. (2004). *First Amendment law in a nutshell* (3rd ed.). St. Paul, MN: Thomson-West.

Beauharnais v. Illinois, 343 U.S. 250 (1952).

Benson, T. W. (1994). Electronic network resources for communication scholars. *Communication Education, 43*(2), 120–128.

Berman, J., & Weitzner, D. J. (1995). Abundance and user control: Renewing the democratic heart of the First Amendment in the age of interactive media. *Yale Law Journal, 104*(7), 1619–1637.

Bostdorff, D. M. (2004). The Internet rhetoric of the Ku Klux Klan: A case study in web site community building run amok. *Communication Studies, 55*(2), 340–361.

Brandenburg v. Ohio, 395 U.S. 444 (1969).

Breckheimer, P. J., II. (2002). A haven for hate: The foreign and domestic implications of protecting Internet hate speech under the First Amendment. *Southern California Law Review, 75*(6), 1493–1528.

Brenner, J. L. (2002). True threats—A more appropriate standard for analyzing First Amendment protection and free speech when violence is perpetrated over the Internet. *North Dakota Law Review, 78*(4), 753–784.

Burch, E. (2001). Censoring hate speech in cyberspace: A new debate in a new America. *North Carolina Journal of Law & Technology, 3*(1), 175–192.

Byassee, W. S. (1995). Jurisdiction of cyberspace: Applying real world precedent to the virtual community. *Wake Forest Law Review, 30*(1), 197–220.

Calvert, C. (1997). Hate speech and its harms: A communication theory perspective. *Journal of Communication, 47*(1), 4–19.

Canis, J. E., & Soriano, E. C. (1996). The Telecommunications Act of 1996: A global analysis. *CommLaw Conspectus, 4*, 147–162.

Cate, F. H. (1995). The First Amendment and the national information infrastructure. *Wake Forest Law Review, 30*(1), 1–50.

Chaplinsky v. New Hampshire, 315 U.S. 568 (1942).

Coglianese, R. N. (1995). Sex, bytes, and community entrapment: The need for a new obscenity standard for the twenty-first century. *Capital University Law Review, 24*(2), 385–422.

Cohen v. California, 403 U.S. 15 (1971).

Collin v. Smith, 578 F.2d 1197 (7th Cir. 1978).

Collins, M. (1994). Internet information management tools. *Communication Education, 43*(2), 112–119.

Cordray, R. (1995). Free speech and the thought we hate. *Ohio Northern University Law Review, 21*(3), 871–895.

Council of Conservative Citizens (CCC). (n.d.). *Council of Conservative Citizens*. Retrieved December 15, 2004, from http://cofcc.org/

Cubby, Inc. v. CompuServe, Inc., 776 F. Supp. 135 (S.D.N.Y. 1991).

Curtis Publishing Co. v. Butts, 388 U.S. 130 (1967).

CyberPatrol. (n.d.). *CyberPatrol Internet security software—Protecting an online generation.* Retrieved May 3 1, 2007, from http://www.cyberpatrol.com/default.aspx?id=123&mnuid=2.5

Cyberspace Communications, Inc. v. Engler, 55 F. Supp. 2d 737 (E.D. Mich. 1999).

Dauterman, W. C., Jr. (2002). Internet regulation: Foreign actors and local harms—At the crossroads of pornography, hate speech, and freedom of expression. *North Carolina Journal of International Law & Commercial Regulation, 28*(1), 177–220.

Debs v. United States, 249 U.S. 211 (1919).

Demaske, C. (2004). Modern power and the First Amendment: Reassessing hate speech. *Communication Law and Policy, 9*(3), 273–316.

Dennis v. United States, 341 U.S. 494 (1951).

Denver Area Educational Telecommunications Consortium, Inc. v. FCC, 518 U.S. 727 (1996).

Dimmick, J. W., Sikand, J., & Patterson, S. J. (1994). The gratifications of the household telephone: Sociability, instrumentality, and reassurance. *Communication Research, 21*(5), 643–663.

Doe v. University of Michigan, 721 F. Supp. 852 (E.D. Mich. 1989).

du Plessis, E. (1994). Recognition versus recall. *Journal of Advertising Research, 34*(3), 75–91.

Epstein, L., & Walker, T. G. (2004). *Constitutional law for a changing America: Rights, liberties, and justice* (5th ed.). Washington, DC: CQ Press.

FCC v. Pacifica Foundation, 438 U.S. 726 (1978).

Fenton, A. T. (1994). Constitutional law—First Amendment—Overbreadth doctrine—Hate crimes—Penalty enhancement. *Duquesne Law Review, 32*(4), 939–956.

Fogo-Schensul, C. (1997/1998). More than a river in Egypt: Holocaust denial, the Internet, and international freedom of expression norms. *Gonzaga Law Review, 33*(1), 241–276.

Frohwerk v. United States, 249 U.S. 204 (1919).

Gey, S. G. (1999). Fear of freedom: The new speech regulation in cyberspace. *Texas Journal of Women & the Law, 8*(2), 183–206.

Gilbert, A. M., & Marchand, E. D. (1999). Splitting the atom or splitting hairs—The Hate Crimes Prevention Act of 1999. *St. Mary's Law Journal, 30*, 931–986.

Gitlow v. New York, 268 U.S. 652 (1925).

Gooding v. Wilson, 405 U.S. 518 (1972).

Griffin, L. M. (1952). The rhetoric of historical movements. *The Quarterly Journal of Speech, 38*(2), 184–188.

Griffin, L. M. (1958). The rhetorical structure of the Antimasonic movement. In D. C. Bryant (Ed.), *The rhetorical idiom: Essays in rhetoric, oratory, language, and drama* (pp. 145–159). Ithaca, NY: Cornell University Press.

Griffin, L. M. (1964). The rhetorical structure of the "New Left" movement: Part I. *The Quarterly Journal of Speech, 50*(2), 113–135.

Haiman, F. S. (1993). *"Speech acts" and the First Amendment.* Carbondale: Southern Illinois University Press.

Hamling v. United States, 418 U.S. 87 (1974).

Harmon, M. D. (1991). Hate groups and cable public access. *Journal of Mass Media Ethics, 6*(3), 146–155.

Hemmer, J. J., Jr. (1986). *The Supreme Court and the First Amendment.* New York: Praeger.

Hess v. Indiana, 414 U.S. 105 (1973).

Jackson, R. L., II., & Heckman, S. M. (2002). Perceptions of white identity and white liability: An analysis of white student responses to a college campus racial hate crime. *Journal of Communication, 52*(2), 434–450.

Jewish Defense League (JDL). (n.d.). *The official Jewish Defense League website.* Retrieved December 16, 2004, from http://www.jdl.org

Komasara, T. (2002). Planting the seeds of hatred: Why imminence should no longer be required to impose liability on Internet communications. *Capital University Law Review, 29*(3), 835–855.

Komsky, S. H. (1991). A profile of users of electronic mail in a university: Frequent versus occasional users. *Management Communication Quarterly, 4*(3), 310–340.

Lange, E. E. (1990). Racist speech on campus: A Title VII solution to a First Amendment problem. *Southern California Law Review, 64*(1), 105–134.

Laquey, T., & Ryer, J. C. (1993). *The Internet companion: A beginner's guide to global networking.* Reading, MA: Addison-Wesley.

Lawley, E. L., & Summerhill, C. (1993). *Internet primer for information professionals: A basic guide to Internet networking technology.* Westport, CT: Mecklermedia.

Lawrence, C. R., III. (1990). If he hollers let him go: Regulating racist speech on campus. *Duke Law Journal, 1990*(3), 431–483.

Leader, S. H., & Probst, P. (2003). The Earth Liberation Front and environmental terrorism. *Terrorism and Political Violence, 15*(4), 37–58.

League of the South (LOS). (n.d.). *League of the South homepage*. Retrieved December 25, 2004, from http://www.dixienet.org/

Leets, L. (2001). Responses to Internet hate sites: Is speech too free in cyberspace? *Communication Law and Policy, 6*(2), 287–317.

Levine, J. R., & Baroudi, C. (1994). *The Internet for dummies* (2nd ed.). San Mateo, CA: IDG Books.

Lipschultz, J. H. (2000). *Free expression in the age of the Internet: Social and legal boundaries*. Boulder, CO: Westview Press.

Macdonald, A. (1980). *The Turner diaries* (2nd ed.). Washington, DC: National Alliance.

Mainstream Loudoun v. Board of Trustees of the Loudoun County Library, 2 F. Supp. 2d 783 (E.D. Va. 1998a).

Mainstream Loudoun v. Board of Trustees of the Loudoun County Library, 24 F. Supp. 2d 552 (E.D. Va. 1998b).

Marsh, E. P. (2000). Purveyors of hate on the Internet: Are we ready for hate spam? *Georgia State University Law Review, 17*(2), 379–407.

Massey, C. R. (1992). Hate speech, cultural diversity, and the foundational paradigms of free expression. *UCLA Law Review, 40*(1), 103–197.

Matsuda, M. J. (1989). Public response to racist speech: Considering the victim's story. *Michigan Law Review, 87*(8), 2320–2381.

McKay, L. J. (1996). The Communications Decency Act: Protecting children from on-line indecency. *Seton Hall Legislative Journal, 20*, 463–502.

Meadows, J. E. (1996). The Telecommunications Act of 1996: Rules of the road for the new highways. *The Computer Lawyer, 13*(3), 1–10.

Melugin v. Hames, 38 F.3d 1478 (9th Cir. 1994).

Miller v. California, 413 U.S. 15 (1973).

Million Youth March, Inc. v. Safir, 63 F. Supp. 2d 381 (S.D.N.Y. 1999).

Mississippi White Knights of the Ku Klux Klan (MWKKKK). (n.d.). Retrieved December 1, 2004, from http://www.mwkkkk.com/

Moore, T. H. (1993). R.A.V. v. City of St. Paul: A curious way to protect free speech. *North Carolina Law Review, 71*(4), 1252–1281.

NAACP v. Claiborne Hardware Co., 458 U.S. 886 (1982).

Nagel, R. F. (1993). Progressive free speech and the uneasy case for campus hate codes. *University of Colorado Law Review, 64*(4), 1055–1059.

Nation of Islam (NOI). (n.d.). *Official Nation of Islam webpage*. Retrieved November 18, 2004, from http://www.noi.org/

National Broadcasting Co. v. United States, 319 U.S. 190 (1943).

National Socialist Movement (NSM). (n.d.). *America's Nazi party: The National Socialist Movement*. Retrieved December 15, 2004, from http://www.nsm88.com/

National Socialist Party of America v. Village of Skokie, 432 U.S. 43 (1977).

New Black Panther Party (NBPP). (n.d.). *Black power!!!!! The New Black Panther Party*. Retrieved November 17, 2004, from http://www.newblackpanther.com/

New York Times v. Sullivan, 376 U.S. 254 (1964).

Novick, S. M. (1989). *Honorable justice: The life of Oliver Wendell Holmes*. Boston: Little, Brown and Company.

O'Callaghan, J. (1994). Free speech by the light of a burning cross. *Cleveland State Law Review, 42*(2), 215–237.

Ogletree, C. J., Jr. (1996/1997). The limits of hate speech: Does race matter? *Gonzaga Law Review, 32*(3), 491–510.

Olson, K. M. (2002). Detecting a common interpretive framework for impersonal violence: The homology in participants' rhetoric on sport hunting, "hate crimes," and stranger rape. *Southern Communication Journal, 67*(3), 215–244.

Orange County Assembly of Christ (OCAC). (n.d.). *OC Assembly of Christ: A Christian Anglo-Israel congregation*. Retrieved December 8, 2004, from http://ocassemblyofchrist.org/

Otte, P. (1994). *The information superhighway: Beyond the Internet*. Indianapolis, IN: Que.

Overbeck, W. (2002). *Major principles of media law* (2002 ed.). Fort Worth, TX: Harcourt.

Pierce v. United States, 252 U.S. 239 (1920).

Planned Parenthood of the Columbia/Williamette v. American Coalition of Life Activists, 290 F.3d 1058 (9th Cir. 2002).

Potok, M. (Ed.). (2004, spring). *Intelligence Report, 113*. (Available from the Southern Poverty Law Center, 400 Washington Avenue, Montgomery, AL 36104)

Potok, M. (Ed.). (2006, spring). *Intelligence Report, 121*. (Available from the Southern Poverty Law Center, 400 Washington Avenue, Montgomery, AL 36104)

Potok, M. (Ed.). (2007, spring). *Intelligence Report, 125*. (Available from the Southern Poverty Law Center, 400 Washington Avenue, Montgomery, AL 36104)

PSINet, Inc. v. Chapman, 108 F. Supp. 2d 611 (W.D. Va. 2000).

R.A.V. v. St. Paul, 505 U.S. 377 (1992).

Red Lion Broadcasting Co. v. FCC, 395 U.S. 367 (1969).

Reed, O. L. (2000). The state is strong but I am weak: Why the "imminent lawless action" standard should not apply to targeted speech that threatens individuals with violence. *American Business Law Journal, 38*(1), 177–208.

Reno v. ACLU, 521 U.S. 844 (1997).

Romero, V. C. (2001). Restricting hate speech against "private figures": Lessons in power-based censorship from defamation law. *Columbia Human Rights Law Review, 33*(1), 1–40.

Roth v. United States, 354 U.S. 476 (1957).

Roy v. United States, 416 F.2d 874 (9th Cir. 1969).

Sable Communications of California v. FCC, 492 U.S. 115 (1989).

Salimipour, N. (2001–2002). The challenge of regulating hate and offensive speech on the Internet. *Southwestern Journal of Law & Trade in the Americas, 8*(2), 395–421.

Sandmann, W. (1994). Three ifs and a maybe: Mari Matsuda's approach to restricting hate-speech. *Communication Studies, 45*(3/4), 241–262.

Schaefer v. United States, 251 U.S. 466 (1920).

Schenck v. United States, 249 U.S. 47 (1919).

Schlosberg, J. (2001). Judgment on "Nuremberg": An analysis of free speech and anti-abortion threats made on the Internet. *Boston University Journal of Science and Technology Law, 7*(1), 52–79.

Schmitz, J., & Fulk, J. (1991). Organizational colleagues, media richness, and electronic mail: A test of the social influence model of technology use. *Communication Research, 18*(4), 487–523.

Siegel, M. L. (1999). Hate speech, civil rights, and the Internet: The jurisdictional and human rights nightmare. *Albany Law Journal of Science & Technology, 9*(2), 375–398.

Simons, H. W. (1970). Requirements, problems, and strategies: A theory of persuasion for social movements. *The Quarterly Journal of Speech, 56*(1), 1–11.

Simons, H. W. (1991). On the rhetoric of social movements, historical movements, and "top-down" movements: A commentary. *Communication Studies, 42*(1), 94–101.

Smith, C. E. (2002). Intentional infliction of emotional distress: An old arrow targets the new head of the hate hydra. *Denver University Law Review, 80*(1), 1–61.

Smith, R. R. (1980). The historical criticism of social movements. *Central States Speech Journal, 31*(4), 290–297.

Smith, S. A. (1994). Communication and the Constitution in cyberspace. *Communication Education, 43*(2), 87–101.

Southeast Booksellers Association v. McMaster, 282 F. Supp. 2d 389 (S.C. 2003).

Southern Poverty Law Center. (1999). *Hate on the net: As new sites arise, so do legal issues*. Retrieved May 24, 2007, from http://www.splcenter.org/intel/intelreport/article.jsp?pid=629

Southern Poverty Law Center. (2003, March). *Hate group numbers rise, but neo-Nazis in disarray*. Retrieved May 24, 2007, from http://www.splcenter.org/center/splcreport/article.jsp?aid=19

Spector, J. (2002). Spreading angst or promoting free expression? Regulating hate speech on the Internet. *University of Miami International and Comparative Law Review, 10*, 155–177.

Spence v. Washington, 418 U.S. 405 (1974).

Stewart, C. J. (1980). A functional approach to the rhetoric of social movements. *Central States Speech Journal, 31*(4), 298–305.

Stromberg v. California, 283 U.S. 359 (1931).

Strossen, N. (1990). Regulating racist speech on campus: A modest proposal? *Duke Law Journal, 1990*(3), 484–573.

Strossen, N. (2001). Incitement to hatred: Should there be a limit? *Southern Illinois University Law Journal, 25*, 243–280.

Terminiello v. Chicago, 337 U.S. 1 (1949).

Texas League of the South. (n.d.). *League of the South membership application page.* Retrieved May 22, 2004, from http://www.texasls.org/join/application.shtml

Texas v. Johnson, 491 U.S. 397 (1989).

*The threat of eco-terrorism: Hearing before the Subcommittee on Forests and Forest Health of the House Resources Committee, House of Representatives,* 107th Cong., 1 (2002).

Timofeeva, Y. A. (2003). Hate speech online: Restricted or protected? Comparison of regulations in the United States and Germany. *Journal of Transnational Law & Policy, 12*(2), 253–285.

Tinker v. Des Moines Independent Community School District, 393 U.S. 503 (1969).

Tsesis, A. (2000). The empirical shortcomings of First Amendment jurisprudence: A historical perspective on the power of hate speech. *Santa Clara Law Review, 40*(3), 729–786.

Tsesis, A. (2001). Hate in cyberspace: Regulating hate speech on the Internet. *San Diego Law Review, 38*(3), 817–874.

Tsesis, A. (2002). Prohibiting incitement on the Internet. *Virginia Journal of Law and Technology, 7*(2), 1–41.

Tualatin Valley Skins (TVS). (n.d.). *88 Aryan skins.* Retrieved November 24, 2004, from http://www.nukeisrael.com/

United States v. Alaboud, 347 F.3d 1293 (11th Cir. 2003).

United States v. Alkhabaz, 104 F.3d 1492 (6th Cir. 1997).

United States v. Baker, 890 F. Supp. 1375 (E.D. Mich. 1995).

United States v. Callahan, 702 F.2d 964 (11th Cir. 1983).

United States v. Darby, 37 F.3d 1059 (4th Cir. 1994).

United States v. Eichman, 496 U.S. 310 (1990).

United States v. Gilbert, 884 F.2d 454 (9th Cir. 1989).

United States v. Kelner, 534 F.2d 1020 (2d Cir. 1976).

United States v. Kosma, 951 F.2d 549 (3d Cir. 1991).

United States v. Malik, 16 F.3d 45 (2d Cir. 1994).

United States v. Morales, 272 F.3d 284 (5th Cir. 2001).

United States v. O'Brien, 391 U.S. 367 (1968).

United States v. Orozco-Santillan, 903 F.2d 1262 (9th Cir. 1990).

United States v. Whiffen, 121 F.3d 18 (1st Cir. 1997).

UWM Post v. Board of Regents of the University of Wisconsin System, 774 F. Supp. 1163 (E.D. Wis. 1991).

Village of Skokie v. National Socialist Party of America, 69 Ill.2d 605 (1978).

Virginia v. Black, 538 U.S. 343 (2003).

Ward v. Rock Against Racism, 491 U.S. 781 (1989).

Watts v. United States, 394 U.S. 705 (1969).

Weintraub-Reiter, R. (1998). Hate speech over the Internet: A traditional constitutional analysis or a new cyber constitution? *Boston University Public Interest Law Journal, 8*(1), 145–173.

Whitney v. California, 274 U.S. 357 (1927).

Wisconsin v. Mitchell, 508 U.S. 476 (1993).

Wood v. Georgia, 370 U.S. 375 (1962).

Yeazel, B. J. (2002). Bomb-making manuals on the Internet: Maneuvering a solution through First Amendment jurisprudence. *Notre Dame Journal of Law, Ethics & Public Policy, 16*(1), 279–314.

Zarefsky, D. (1980). A skeptical view of movement studies. *Central States Speech Journal, 31*(4), 245–254.

# Index